Reviews of Environmental Contamination and Toxicology

VOLUME 198

Reviews of Environmental Contamination and Toxicology

Editor
David M. Whitacre

VOLUME 198

ISBN: 978-1-4419-1878-9 e-ISBN: 978-0-387-09647-6
DOI: 10.1007/978-0-387-09647-6

Printed on acid-free paper

9 8 7 6 5 4 3 2 1

springer.com

Foreword

International concern in scientific, industrial, and governmental communities over traces of xenobiotics in foods and in both abiotic and biotic environments has justified the present triumvirate of specialized publications in this field: comprehensive reviews, rapidly published research papers and progress reports, and archival documentations. These three international publications are integrated and scheduled to provide the coherency essential for nonduplicative and current progress in a field as dynamic and complex as environmental contamination and toxicology. This series is reserved exclusively for the diversified literature on "toxic" chemicals in our food, our feeds, our homes, recreational and working surroundings, our domestic animals, our wildlife and ourselves. Tremendous efforts worldwide have been mobilized to evaluate the nature, presence, magnitude, fate, and toxicology of the chemicals loosed upon the earth. Among the sequelae of this broad new emphasis is an undeniable need for an articulated set of authoritative publications, where one can find the latest important world literature produced by these emerging areas of science together with documentation of pertinent ancillary legislation.

Research directors and legislative or administrative advisers do not have the time to scan the escalating number of technical publications that may contain articles important to current responsibility. Rather, these individuals need the background provided by detailed reviews and the assurance that the latest information is made available to them, all with minimal literature searching. Similarly, the scientist assigned or attracted to a new problem is required to glean all literature pertinent to the task, to publish new developments or important new experimental details quickly, to inform others of findings that might alter their own efforts, and eventually to publish all his/her supporting data and conclusions for archival purposes. In the fields of environmental contamination and toxicology, the sum of these concerns and responsibilities is decisively addressed by the uniform, encompassing, and timely publication format of the Springer triumvirate:

Reviews of Environmental Contamination and Toxicology [Vol. 1 through 97 (1962–1986) as Residue Reviews] for detailed review articles concerned with any aspects of chemical contaminants, including pesticides, in the total environment with toxicological considerations and consequences.

Bulletin of Environmental Contamination and Toxicology (Vol. 1 in 1966) for rapid publication of short reports of significant advances and discoveries in the fields of air, soil, water, and food contamination and pollution as well as methodology and other disciplines concerned with the introduction, presence, and effects of toxicants in the total environment.

Archives of Environmental Contamination and Toxicology (Vol. 1 in 1973) for important complete articles emphasizing and describing original experimental or theoretical research work pertaining to the scientific aspects of chemical contaminants in the environment.

Manuscripts for *Reviews* and the *Archives* are in identical formats and are peer reviewed by scientists in the field for adequacy and value; manuscripts for the *Bulletin* are also reviewed, but are published by photo-offset from camera-ready copy to provide the latest results with minimum delay. The individual editors of these three publications comprise the joint Coordinating Board of Editors with referral within the Board of manuscripts submitted to one publication but deemed by major emphasis or length more suitable for one of the others.

Coordinating Board of Editors

Preface

The role of *Reviews* is to publish detailed scientific review articles on all aspects of environmental contamination and associated toxicological consequences. Such articles facilitate the often-complex task of accessing and interpreting cogent scientific data within the confines of one or more closely related research fields.

In the nearly 50 yr since *Reviews of Environmental Contamination and Toxicology* (formerly *Residue Reviews*) was first published, the number, scope and complexity of environmental pollution incidents have grown unabated. During this entire period, the emphasis has been on publishing articles that address the presence and toxicity of environmental contaminants. New research is published each yr on a myriad of environmental pollution issues facing peoples worldwide. This fact, and the routine discovery and reporting of new environmental contamination cases, creates an increasingly important function for *Reviews*.

The staggering volume of scientific literature demands remedy by which data can be synthesized and made available to readers in an abridged form. *Reviews* addresses this need and provides detailed reviews worldwide to key scientists and science or policy administrators, whether employed by government, universities or the private sector.

There is a panoply of environmental issues and concerns on which many scientists have focused their research in past yr. The scope of this list is quite broad, encompassing environmental events globally that affect marine and terrestrial ecosystems; biotic and abiotic environments; impacts on plants, humans and wildlife; and pollutants, both chemical and radioactive; as well as the ravages of environmental disease in virtually all environmental media (soil, water, air). New or enhanced safety and environmental concerns have emerged in the last decade to be added to incidents covered by the media, studied by scientists, and addressed by governmental and private institutions. Among these are events so striking that they are creating a paradigm shift. Two in particular are at the center of ever-increasing media as well as scientific attention: bioterrorism and global warming. Unfortunately, these very worrisome issues are now super-imposed on the already extensive list of ongoing environmental challenges.

The ultimate role of publishing scientific research is to enhance understanding of the environment in ways that allow the public to be better informed. The term "informed public" as used by Thomas Jefferson in the age of enlightenment conveyed the thought of soundness and good judgment. In the modern sense, being "well informed" has the narrower meaning of having access to sufficient information. Because the public still gets most of its information on science and technology from TV news and reports, the role for scientists as interpreters and brokers of scientific information to the public will grow rather than diminish.

Environmentalism is the newest global political force, resulting in the emergence of multi-national consortia to control pollution and the evolution of the environmental ethic. Will the new politics of the 21st century involve a consortium of technologists and environmentalists, or a progressive confrontation? These matters are of genuine concern to governmental agencies and legislative bodies around the world.

For those who make the decisions about how our planet is managed, there is an ongoing need for continual surveillance and intelligent controls, to avoid endangering the environment, public health, and wildlife. Ensuring safety-in-use of the many chemicals involved in our highly industrialized culture is a dynamic challenge, for the old, established materials are continually being displaced by newly developed molecules more acceptable to federal and state regulatory agencies, public health officials, and environmentalists.

Reviews publishes synoptic articles designed to treat the presence, fate, and, if possible, the safety of xenobiotics in any segment of the environment. These reviews can either be general or specific, but properly lie in the domains of analytical chemistry and its methodology, biochemistry, human and animal medicine, legislation, pharmacology, physiology, toxicology and regulation. Certain affairs in food technology concerned specifically with pesticide and other food-additive problems may also be appropriate.

Because manuscripts are published in the order in which they are received in final form, it may seem that some important aspects have been neglected at times. However, these apparent omissions are recognized, and pertinent manuscripts are likely in preparation or planned. The field is so very large and the interests in it are so varied that the Editor and the Editorial Board earnestly solicit authors and suggestions of underrepresented topics to make this international book series yet more useful and worthwhile.

Justification for the preparation of any review for this book series is that it deals with some aspect of the many real problems arising from the presence of foreign chemicals in our surroundings. Thus, manuscripts may encompass case studies from any country. Food additives, including pesticides, or their metabolites that may persist into human food and animal feeds are within this scope. Additionally, chemical contamination in any manner of air, water, soil, or plant or animal life is within these objectives and their purview.

Manuscripts are often contributed by invitation. However, nominations for new topics or topics in areas that are rapidly advancing are welcome. Preliminary communication with the Editor is recommended before volunteered review manuscripts are submitted.

Summerfield, North Carolina D.M.W.

Contents

Foreword v

Preface vii

1 **The Impact of Environmental Chemicals on Wildlife Vertebrates** 1
Julia Bernanke and Heinz-R. Köhler

2 **Biomarkers in Aquatic Plants: Selection and Utility** 49
Richard A. Brain and Nina Cedergreen

3 **Human Health Effects of Methylmercury Exposure** 111
Sergi Díez

4 **Waterborne Adenovirus** 135
Kristina D. Mena and Charles P. Gerba

5 **Haloacetonitriles: Metabolism and Toxicity** 171
John C. Lipscomb, Ebtehal El-Demerdash and Ahmed E. Ahmed

Index 201

The Impact of Environmental Chemicals on Wildlife Vertebrates

Julia Bernanke and Heinz-R. Köhler (✉)

Contents

1 Introduction 2
2 The Impact of Anthropogenic Chemicals on Fish 4
2.1 Reproductive Parameters in Osteichthyes 4
2.2 Induction of Vitellogenin 6
2.3 Abnormal Gonadal Development 7
2.4 The Impact of Pulp and Paper Mill Effluents 8
2.5 The Impact of Heavy Metals 9
2.6 Reproductive Parameters in Chondrichthyes 10
2.7 Impact on Fish: Conclusion 11
3 The Impact of Anthropogenic Chemicals on Amphibians 12
3.1 Reproductive and Developmental Parameters in Amphibians 13
3.2 Vulnerability of Early Life Stages 13
3.3 The Impact of Pesticides and PCBs 14
3.4 The Impact of Fertilizers 17
3.5 Deformities in Amphibians 18
3.6 Impact on Amphibians: Conclusion 18
4 The Impact of Anthropogenic Chemicals on Reptiles 19
4.1 The Impact of EDCs on Gender Determination 20
4.2 The Impact of EDCs on Reptile Populations 21
4.3 Developmental Abnormalities in Reptiles 22
4.4 Impact on Reptiles: Conclusion 22
5 The Impact of Anthropogenic Chemicals on Birds 23
5.1 Reproductive Parameters in Birds: Behavior and Sexual Differentiation 24
5.2 The Impact on Reproductive Organs 25
5.3 Great Lakes Embryo Mortality, Edema, and Deformities Syndrome 26
5.4 Altered Sex Skew 26
5.5 Eggshell Thinning 27
5.6 The Impact of Oil Spills 28
5.7 Impact on Birds: Conclusion 29
6 The Impact of Anthropogenic Chemicals on Feral Mammals 30
6.1 Endocrine Disruption in Marine Mammals 31
6.2 Endocrine Disruption in Freshwater Mammals 33

H-R. Köhler
Animal Physiological Ecology, University of Tübingen, D-72072 Tübingen, Germany
heinz-r.koehler@uni-tuebingen.de

D.M. Whitacre (ed.) *Reviews of Environmental Contamination and Toxicology,* Vol 198
doi: 10.1007/978-0-387-09646-9,

6.3 Endocrine Disruption in Terrestrial Mammals 34
6.4 Impact on Feral Mammals: Conclusion 36
7 Summary 36
References 37

1 Introduction

Since early history, humans have interfered with their environment. Early hominids appeared about 6 million yr ago. Although the earliest humans were only able to partially control their environment, they had an impact on nature from their hunting activities. The more severe impact on the environment began later, after the birth of agriculture and particularly after the industrial revolution began.

The tremendous growth of the chemical industry over the last century is a phenomenon of the twentieth and, now the twenty-first century. The increasing production and use of chemicals have reached enormous global dimensions. Some environmentally released chemicals are by-products of manufacturing processes; others are developed for particular applications and are intentionally released to the environment, e.g., pesticide use in agriculture. Occasionally, large quantities of chemicals are released as a result of accidents. Public awareness of risks posed by man made chemicals has grown rapidly over the past few decades, particularly after the release of Rachel Carson's book 'Silent Spring', in 1962. At that time, few chemicals had been well tested for toxic effects on wildlife prior to commercial use. In the succeeding yr, effects were observed on the environment and on wildlife from exposure to various anthropogenic chemicals (Ankley and Giesy 1998; Fox 1992; Van Der Kraak et al. 2001). In fact, the term "environmental chemicals" was coined to describe those chemicals that have a strong impact on the environment, and on humans, animals and plants.

Research undertaken to address the affects of environmental chemicals on wildlife generally entails two complementary approaches (Fig. 1). The first is a "bottom-up" approach; the second is a "top-down approach." In the bottom-up approach, groups of individual organisms are experimentally exposed to chemicals, chemical mixtures or environmental samples, and their effects on health and life cycle (development, fertility or offspring production) are observed and recorded. Undoubtedly, substances that act at specific biological sites, such as endocrine system toxicants, as well as general toxicants can affect organismal longevity, reproduction and other biological processes. However, a major challenge in bottom-up studies is interpreting the potential ecological consequences of observed organismal-level effects (Maltby 1999). When reporting such studies, authors often extrapolate from observations of adverse effects on fertility or reproduction to speculate on alterations in population structure or size. Although such speculation may be imperfect, it is also difficult to establish causes

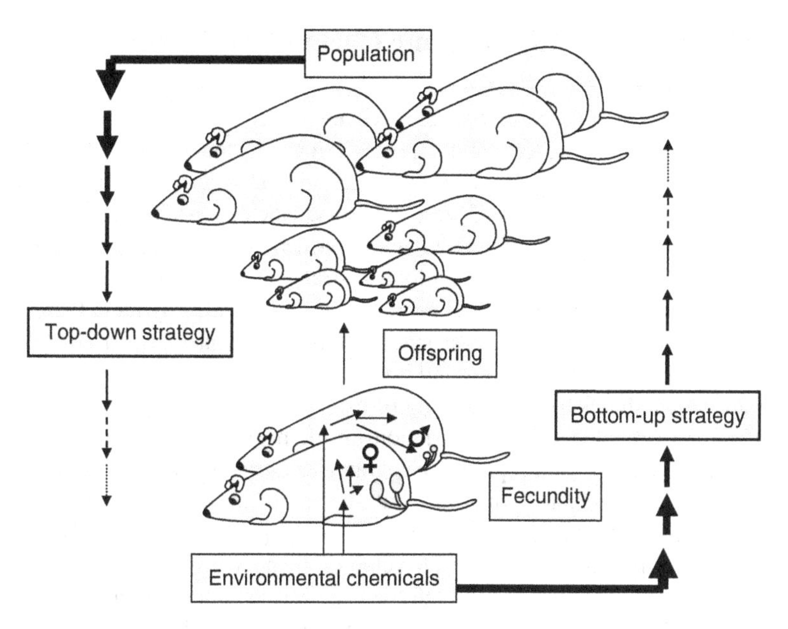

Fig. 1 Depiction of the two strategies used to causally link population effects to chemical pollution. In a "top-down" approach, recorded changes in population dynamics usually are related to effects on reproductive output but, additionally, need to be mechanistically linked to chemical-induced physiological impact on parental fecundity. The alternative is a "bottom-up" approach, which relates the presence and concentration of xenobiotics to reproductive parameters, and tries to extrapolate xenobiotic effect data to the population level of biological organization

of observed changes in population parameters by relying only on field survey data, which constitutes a top-down approach (Barnthouse 1993; Evans et al. 1990). Consequently, such top-down studies on observed population declines sometimes link an observed effect on population structure to a chemical, without considering that changes may result from natural fluctuations in populations, predator-prey interactions, disease outbreaks, or climate and habitat change. Although both bottom-up and top-down studies have their particular justifications, both approaches are needed to gain an integrated view of probable environmental impacts from chemicals and to deal with the normal array of natural complexity and uncertainty.

There is a widely held opinion and assumption that anthropogenically released substances and endocrine disrupting chemicals (EDCs), in particular, have the potential to affect wildlife vertebrate population dynamics. The object of this review is to thoroughly evaluate the literature that underlies this assumption. We have placed emphasis on sifting available scientific evidence to separate fact from speculation, in regard to the chemical-induced population effects of xenobiotic chemicals. To this end, about 250 publications were critically assessed, covering the time span from the late 1940s to the yr 2006.

2 The Impact of Anthropogenic Chemicals on Fish

Seventy-three percent of the earth's surface is covered with water. Vertebrate life began in water and half of living vertebrate species remain in the aquatic environment. Aquatic vertebrates live in oceans, lakes and rivers, in freshwater, brackish and marine environments; osmotic properties of these diverse aquatic ecosystems vary greatly.

Evolution and adaptation to different habitats has led to a high diversity of physiological, anatomical, behavioral and ecological strategies in fish. Fish are known to be the most successful group of vertebrates, and are represented by more than 3,000 species of cartilaginous fish such as elasmobranchs and chimaeras, over 20,000 species of bony fish (teleosts, dipnoans and holosteans), and several species of jawless fish (lampreys and hagfish). The diversity of fish and the habitats in which they live have offered unparalleled scope for variations of life history. For example, fish have developed various successful reproductive strategies during evolution (Kime 1998).

The majority of fish are oviparous, but species displaying ovoviviparity and viviparity are also widespread. Most oviparous species produce millions of eggs that are released into the water to drift and develop on their own; other species produce fewer, though larger eggs and guard both the eggs and the young. In addition, some fish species display sexual plasticity, a reproductive strategy, which is rather uncommon among vertebrates. Further, it is known that certain teleosts, particularly coral reef fishes can change sex in response to environmental changes, during their lifetimes.

The heterogeneity in physiology, anatomy, behavior and ecology of various fish species makes evaluating possible affects from anthropogenic chemical exposure difficult. Van der Kraak et al. (2001) suggested that aquatic respiration may render fish more susceptible to increased chemical exposure. Chemical compounds easily enter the body via the gills and rapidly enter the bloodstream. Osmoregulation plays an important role in the sensitivity of fish to environmental chemicals; marine teleosts that drink seawater may be particularly sensitive to environmental chemicals. In contrast, hyperosmotic freshwater fishes are subjected to a steady influx of water into their bodies, which offers another possible entry pathway for chemicals.

As in other vertebrate groups (e.g., mammals), fish may also store hydrophobic xenobiotic contaminants in their adipose tissue. Such lipid-soluble compounds may affect fish, particularly during critical periods of development. Van der Kraak et al. (2001) suggested that a release of lipid-soluble compounds from adipose tissue has the potential to harm the physiology of fish more seriously than the levels accumulated directly from the water.

2.1 *Reproductive Parameters in Osteichthyes*

Chemicals may impact the endocrine system in fish in diverse ways as we will discuss in the following chapters. However, few data are available on the impact of environmental chemicals on freshwater fish populations (Jobling and Tyler 2003).

Jobling and Tyler (2003) studied lake trout (*Salvelinus namaycush*) from Lake Ontario, Canada/USA, and linked their recent decline to organochlorine chemical exposure; organochlorines are known to have endocrine disrupting effects. In addition, the reproductive systems of several of the following flatfish populations were affected by environmental chemicals: winter flounder (*Pleuronectes americanus*) (Johnson et al. 1992), rock sole (*Pleuronectes bilineatus*) (Johnson et al. 1998), and English sole (*Parophrys vetulus*) (Casillas et al. 1991). All authors observed and reported inhibited ovarian development, reduced larval survival, reduced egg weight, precocious female maturation and reduced spawning success. Similar results were observed in Atlantic herring (*Clupea harengus*) (Hansen et al. 1985), Atlantic cod (*Gadus morhua*) (Petersen et al. 1997), and European flounder (Von Westernhagen et al. 1981) from the Baltic Sea, and also in lake trout (*S. namaycush*) from the Great Lakes, Canada /USA (Mac et al. 1993), and Arctic char (*Salvelinus alpinus*) from Lake Geneva in Switzerland/France (Monod 1985). Although the observed effects are real, the mechanisms behind them are not well understood. A correlation was detected between the observed effects and increased levels of polyaromatic hydrocarbons (PAHs), dichloro-diphenyl-trichloroethane (DDT) and polychlorinated biphenyls (PCBs) in exposed animals. The majority of the effects, especially reduced fecundity and spawning success, may be linked to the anti-estrogenic effects of some PCBs and PAHs (Casillas et al. 1991; Johnson et al. 1992, 1998).

Oil spills are occurring more frequently worldwide. It is consequently expected that affects on fish populations from contact with oil will increase. The endocrine system in fish may be affected by oil spills, and may lead to population-wide effects in some species.

Following the Amoco Cadiz oil spill in Brittany, France in March 1978, nearly 223,000 t of oil was released into the sea. First, it was suggested that the oil had relatively little impact on fish species; further research showed that fish species had been affected. Oil exposure resulted in several observed effects, including reduced and delayed ovarian development, tumors and ulcerations in plaice (*Pleuronectes platessa*) (Stott et al. 1983).

After the Exxon Valdez oil spill in Prince William Sound, Alaska, in 1989, resultant effects on fish populations were examined; the status of the wildstock pink salmon (*Oncorhynchus gorbuscha*) population decreased dramatically. However, in the yr following the spill concentrations of PAHs in the water declined and fish populations were observed to rebound (Maki et al. 1995). Hilborn and Eggers (2000) suggested that the observed decline in wild stocks of pink salmon began well before the oil spill occurred, and that the oil spill accident did not exacerbate the population decline. In contrast, the Pacific herring (*Clupea pallasi*) population in Prince William Sound collapsed after the oil spill. It was concluded that both the oil spill and over fishing contributed to this population collapse (Thorne 2004). Although the population recovered, it again suffered a decline in 1992–1993. Carls et al. (2002) suggested that this second collapse was caused by high population size and diseases, although lingering effects from the oil spill could not be ruled out.

In addition to the studies described above, it was reported that fish populations (mainly brown trout, *Salmo trutta*) in many Swiss rivers and streams declined dramatically over the last 15 yr. To evaluate the causes for the observed decline, the project 'Fishnet' was established in 1998. However, the project failed to unequivocally discover why these Swiss fish populations declined. The authors cited the following factors as contributors to the decline: reproductive failure, habitat degradation, reduced quality of habitat, and climatic changes (such as increased water temperature and shifts in the seasonal occurrence of floods) (Burkhardt-Holm et al. 2002).

2.2 Induction of Vitellogenin

It has been observed that the alteration of sex hormones in fish may lead to reproduction changes or failure (Kime et al. 1999; Tyler and Routledge 1998). There are many EDCs (e.g., PCBs or PAHs) that potentially mimic natural sex hormones in organisms. One well-investigated example is the induction of vitellogenin caused by EDCs. Normally, vitellogenin is only produced subsequent to the binding of estradiol, a typical female sex hormone, to its receptor protein and, therefore, the production of vitellogenin occurs typically only in females.

Tyler and Routledge (1998) suggested that the stimulation of vitellogenin in male and juvenile fish can be used as a biomarker to identify environmental estrogen and estrogenic chemical effects. Furthermore, Kime et al. (1999) believed that perturbations of female vitellogenin levels could provide a useful marker of endocrine-induced dysfunction. Therefore, the effect of potentially-disruptive estrogenic substances may be detected by tracking the concentration of vitellogenin in both male and female fishes. Currently, research is being focused on vitellogenin induction in fish in a several water bodies globally. Fish with abnormal vitellogenin concentrations were found in the USA, Sweden, UK, the Netherlands, Canada, Germany, Switzerland, and Japan. Dramatic increases in vitellogenin levels have been observed, particularly in freshwater and marine fish exposed to sewage treatment works.

Examples with the clearest effects come from the United Kingdom (UK). Purdom et al. (1994) and Harries et al. (1996, 1997) evaluated vitellogenin induction in male rainbow trout (*Oncorhynchus mykiss*). These fish were caught in highly polluted rivers and showed increased plasma vitellogenin levels. The elevated plasma vitellogenin levels detected in these trout appeared to result from high concentrations of estrogenic compounds in the rivers from which they came. Purdom et al. (1994) also observed high vitellogenin levels in immature carp (*Cyprinus carpio*) from the UK.

In another study, male carp (*C. carpio*) exposed to sewage treatment discharges from different localities in the USA showed elevated plasma vitellogenin levels (Folmar et al. 1996).

Similar results were seen with European flounder (*P. flesus*) caught in heavily polluted estuaries (Tyne, Tees, Wear and Mersey estuaries) in the UK. Plasma vitellogenin levels up to 20 mg/ml plasma were observed (Matthiessen and Gibbs 1998). Again, it was hypothesized that estrogenic discharges were responsible for the observed effects. This suspicion was confirmed when male flounder (*P. flesus*), from less polluted waters, were found to have low or undetectable levels of vitellogenin. Fent (2003) also discovered that the degree of vitellogenesis in fish declined as distance from the source of pollution increased.

Japanese flounder (*Pleuronectes yokohamae*) from Tokyo Bay, Japan showed increased vitellogenin levels as well. Although it was not possible to detect the causative substances for the observed effect, it was assumed that estrogenic chemicals found in the estuary were responsible (Hashimoto et al. 2000).

Nichols et al. (1999) examined fathead minnows (*Pimephales promelas*) exposed to municipal wastewaters at different localities in the USA, and the observed fish showed vitellogenin induction. There is fear that paper mill effluents may similarly affect fish exposed to them. A study by Mellanen et al. (1999) confirmed the suspicion: whitefish (*Coregonus lavaretus*) exposed to pulp and paper mill efflux showed elevated vitellogenin levels.

In summary, numerous authors have observed that EDCs have the potential to mimic natural hormones. The induction of vitellogenin by EDCs has occurred and is occurring in many species globally. However, the consequences of vitellogenin induction on fish reproduction and population dynamics are unknown.

2.3 Abnormal Gonadal Development

It is generally believed that environmental chemicals such as 17β-estradiol or alkylphenols, found in discharges of sewage treatment works or pulp and paper mill effluents, not only cause vitellogenin induction but also lead to reduced testicular growth, enlarged livers or altered gonadal development in exposed fish. Vitellogenin induction and testicular abnormalities were detected in male European flounder (*P. flesus*) that were caught near a sewage treatment plant discharge in England (Lye et al. 1997, 1998). It was hypothesized that the observed effects resulted from exposure to, and bioaccumulation of, several estrogenic alkylphenols (Lye et al. 1999). A study by Jobling et al. (1996) confirmed the suspicion that estrogenic compounds can cause inhibition of testicular growth. Rainbow trout (*O. mykiss*) were exposed to realistic alkylphenol concentrations under laboratory conditions; results showed effects similar to those observed in nature. Harries et al. (1997) also observed the inhibition of testicular growth in adult rainbow trout (*O. mykiss*) exposed to water heavily contaminated with estrogenic compounds (alkylphenols).

Gill et al. (2002) studied testicular development and spermatogenesis in male European flounder (*P. flesus*). The fish were taken from the lower Tyne estuary in northeast England, a heavily contaminated site, known to contain high levels of EDCs. Abnormal changes in testicular structure were detected. Most notable was hypertrophy of connective tissue. Gill et al. (2002) further suggested that the incidence of sperm and testicular abnormalities observed will impact the reproductive success of the flounder.

Wild male roach (*Rutilus rutilus*) and gudgeon (*Gobio gobio*) collected downstream of a sewage treatment facility that discharges into several rivers in the UK showed the presence of ovotestes. However, it was observed that some fish had only an occasional oocyte in otherwise normal testicular tissue, while other fish suffered from large regions of abnormal testicular tissue (Jobling et al. 1998a). Estrogenic chemicals (particularly 17β-estradiol) were said to be responsible for the observed effects (Tyler and Routledge 1998). Exposure to these chemicals may lead to other testicular abnormalities such as feminized or absent vasa deferentia and impaired milt production (Jobling et al. 1998b). Altered spermatogenesis was found in wild European flounder (*P. flesus*) from heavily industrialized estuaries in England and France (Lye et al. 1998). Various authors reported that exposed flounder showed ovotestis (Allen et al. 1999; Minier et al. 2000).

Cases of possible but unconfirmed endocrine disruption-related effects on gonad development by potential EDCs are numerous; such effects include decreased egg weight and increased atresia of oocytes in flatfish (*Parophrys vetulus, Lepidopsetta bilineat,* and *Platichthys stellatus*) from contaminated harbors in the eastern USA (Johnson et al. 1992), and premature maturation in flatfish from the southern portion of the North Sea (Rijnsdorp and Vethaak 1997).

Many studies have focused on the possible causes for the disturbed gonadal development and abnormal testicular tissue in fish. It was observed that effluents from sewage treatment facilities have the potential to affect exposed fish. It was further observed that those effluents contain EDCs, which were thought to be responsible for the observed alterations.

2.4 The Impact of Pulp and Paper Mill Effluents

Gagnon et al. (1995) reported reproductive abnormalities in fish exposed to pulp and paper mill effluents. Those effluents contain chemicals with endocrine disrupting properties that masculinize females. This in turn may result in suppressed male and female reproduction, reduced gonad size, and more variable fecundity. However, the active compounds responsible for the reproductive abnormalities in exposed fish have not yet been identified. The observed reproductive abnormalities induced by pulp and paper mill effluents are very similar among many fish species. Fentress et al. (2006) evaluated the hormonal status of wild longear sunfish (*Lepomis megalotis*) in a US river (Pearl River at Bogalusa, LA) receiving unbleached kraft and recycled pulp mill effluent. The effluent from this mill was found to suppress female testosterone and vitellogenin levels, when it constituted more than 1% of river flow. Reproductive suppression was observed in longear sunfish, in response to contact with unbleached kraft and recycled pulp mill effluent, but it remained unclear whether this effect was reflected in the population structure.

Munkittrick et al. (1991, 1998) observed that longnose sucker (*Catostomus catostomus*) and lake whitefish (*Coregonus clupeaformis*) exposed to pulp and paper mill effluents in Canada exhibited reduced gonadal size and delayed sexual maturity.

The authors concluded that estrogenic chemicals in the effluent may have profound impact on the reproduction of exposed fish, despite a lack of direct proof. Andersson et al. (1988) and Sandström et al. (1988) reported similar results for Eurasian perch (*Perca fluviatilis*) and blenny (*Zoarces viviparous*) exposed to pulp and paper mill effluents in Scandinavia. Furthermore, it was shown that effects on fish decrease, with distance, downstream of pulp and paper mills. In two additional studies, Munkittrick et al. (1992a, b) observed reproductive abnormalities and delayed sexual maturation in lake whitefish (*C. clupeaformis*) that inhabit an area of Jackfish Bay, Canada, where mill effluents have caused serious chemical contamination.

In addition to the foregoing examples, it was observed that exposure of phytoestrogens and bleached kraft mill effluents may lead to reduced levels of sex steroids in both male and female fish (Robinson et al. 1994; Kovacs et al. 1995). Robinson et al. (1994) and Kovacs et al. (1995) conducted laboratory experiments on sex steroid levels in fathead minnows (*P. promelas*) exposed to bleached kraft-mill effluents. The authors observed that altered sex steroid levels profoundly affect reproduction in fathead minnows. Depressed hormonal levels were also found in white suckers (*C. commersoni*) in the USA (Hodson et al. 1992) and longnose sucker (*C. catostomus*) collected at Jackfish Bay, Canada (Munkittrick et al. 1992a, b). Both species were exposed to bleached kraft mill effluents.

The foregoing studies provide evidence that reproductive responses were directly associated with effluent exposure, rather than resulting from other environmental factors such as habitat alteration. The accumulated evidence indicates that the observed reproductive effects and hormonal changes in fish result from contact with constituents of pulp and paper mill effluents, including those that disrupt endocrine function.

2.5 The Impact of Heavy Metals

The impact that heavy metals may have on fish populations is best illustrated by the copper redhorse (*Moxostoma hubbsi*), an endangered fish species the worldwide distribution of which is limited to the St. Lawrence River and three of its Canadian tributaries. De Lafontaine et al. (2002) observed accidentally killed individuals and found high concentrations of total mercury (Hg), cadmium (Cd), and co-planar PCBs in them. Although it was not possible to prove the link between the observed contaminants and reproductive failure in this endangered fish species, PCBs appeared to be the culprits responsible for the decline of the fish population (De Lafontaine et al. 2002).

In another study from North-eastern Ontario, Canada, 18 lakes were studied over a 5-yr period to evaluate effects of Cd, copper (Cu) and other heavy metals on yellow perch (*Perca flavescens*). The results indicated that chronic metal exposure of fish may lead to impaired aerobic capacities, altered aerobic swim performance and respiration rate in wild yellow perch. The authors believed that metal contamination can affect health and may alter population dynamics of yellow perch (Couture and Rajotte 2003).

Much attention has been paid to the bull trout (*Salvelinus confluentus*), which was recently listed as a threatened organism in the U. S. Federal Endangered Species Act. Hansen et al. (2002) discussed the possible threats to this species. Past and present habitat for the bull trout includes waterways contaminated with heavy metals released from mining activities. The authors suggested that the sensitivity of bull trout to Cu was one possible reason for the population decline.

Alquezar et al. (2006) compared the condition and reproductive output of toadfish (*Tetractenos glaber*) in metal contaminated (and reference) estuaries near Sydney, Australia. A positive relationship was observed in this species, between levels of lead (Pb) and decreased oocyte diameter and density. The results suggested a possible decline in female reproductive output caused by reduction in egg size and fecundity. This, in turn, may affect fish population and community structure (Alquezar et al. 2006).

Cd, Hg and Pb are all suspected of having endocrine disrupting properties. Nevertheless, distinguishing between hormone-induced effects and impairment of reproductive parameters resulting from under-nutrition is difficult.

2.6 Reproductive Parameters in Chondrichthyes

Only limited data are available on chemical affects to cartilaginous fish.

Chondrichthyan species are not as productive as are bony fishes, a consequence of their different life-history strategies, which renders them more vulnerable to environmental impacts. Indeed, over fishing or by-catches may be more threatening to cartilaginous fish than exposure to anthropogenic chemicals. In Canada, blue sharks (*Prionace glauca*) are the ones most commonly caught. Blue sharks have been in steady decline during recent yr because of high international catch mortality (Campana et al. 2006). The declining elasmobranch populations, within the United Kingdom's coastal zone, are thought to result from development of installations that generate offshore renewable energy. Yet clear causes for population decline are not known, although interactions with wind farms may be one reason (Gill and Kimber 2005). A few studies have described a possible link between anthropogenic chemicals and the decline of elasmobranches.

Because of their persistence in aquatic environments and ability to impair reproduction and other critical physiological processes, organochlorine contaminants pose significant health risks to marine organisms. Despite such concerns, few studies have been undertaken to investigate the degree to which sharks are exposed to organochlorines. These fish are easily threatened by anthropogenic pollution because of their tendency to excessively bioaccumulate and biomagnify environmental contaminants. Gelsleichter et al. (2005) examined concentrations of organochlorine pesticides and PCBs in the bonehead shark (*Sphyrna tiburo*) from four estuaries on Florida's Gulf coast, Apalachicola Bay, Tampa Bay, Florida Bay and Charlotte Harbor in the USA.They found that organochlorine concentrations in *S. tiburo* were higher in Apalachicola Bay, Tampa Bay, and Charlotte Harbor than in the Florida Bay population. Because the rate of infertility was dramatically higher for *S. tiburo,* in Tampa Bay than in Florida Bay, the present findings allude to a possible relationship between organochlorine exposure and reproductive health (Gelsleichter et al. 2005).

The effect of tributyltin oxide (TBTO), the main constituent of tin-based antifouling marine paint, was observed in another study on stingrays (*Urolophus jamaicensis*). Tin accumulated in the gill tissue of the stingray after acute exposure to TBTO. Results of this study included: alterations in the morphological architecture of the gill, induction of stress proteins and peroxidative damage, in response to tributyltin (TBT) exposure. However, neither reproductive parameters, nor population structure were affected (Dwivedi and Trombetta 2006).

Concentrations of PCBs and organochlorine pesticides (DDTs) were measured in the liver of two shark species, blue shark (*P. glauca*) and kitefin shark (*Dalatias licha*), from the Mediterranean Sea. Shark tissues were highly contaminated, which suggests that organochlorine pesticide contamination still exists in this marine environment, and may give rise to future population effects (Storelli et al. 2005).

Although only limited data are available on the possible impacts of environmental compounds on chondrichtyes, organochlorines may also induce effects in this class. However, a clear link between these chemicals and possible impacts on population levels of Chondrichtyes is not yet supported in the literature.

2.7 *Impact on Fish: Conclusion*

Fish populations, mainly freshwater fish populations, are particularly sensitive to anthropogenic chemicals. Their aquatic life may place them in constant exposure to chemicals with hormone-like properties. Uptake of chemicals such as PCBs or PAHs readily occurs via the gills and skin, as well as via the diet.

Endocrine disruption in wild fish has been observed in America, Asia, Australia and Europe. Despite wide-spread reports of endocrine disruption in fish, especially in teleosts (and this is well-characterized at the species level), few studies have demonstrated population-level consequences as a result of exposure to EDCs (Van der Kraak et al. 2001).

There is compelling global evidence that exposure to EDCs is compromising the physiology and sexual behavior of fish, including effecting permanent alteration of sexual differentiation and impairment of fertility.

Johnson et al. (1997), believed that non-endocrine causes for affects on fish should also be evaluated because of the large number of contaminants that exist in the environment. In a 5-yr study, Triebskorn et al. (2001) have combined active and passive biomonitoring experiments on brown trout (*Salmo trutta* f. *fario*) and stone loach (*Barbatula barbatula*) with laboratory studies. This consortium investigated molecular, cellular, physiological, developmental, reproductive, and ecological markers at both population and community levels (also see references in Triebskorn and Köhler (2001)). Using Hill's plausibility criteria, in a weight of evidence approach, it was possible to link observed subcellular, individual, and ecological effects (Triebskorn et al. 2003), and to assemble a clear picture of chemical impacts to different levels of biological organization, including the population level. Despite serious attempts, however, the mechanistic pathways between the biological levels could not be determined in detail.The population decline of various Chondrichtyes species appears to result from over-fishing or by-catching more than from the impact of anthropogenic chemicals. Almost all cartilaginous fish species are generally slow growing, late maturing, and produce relatively few young. Therefore, they are vulnerable to population impacts exceeding 2–3% loss of population numbers, for example, by fishing, or from other reasons (Abbott 2000; http://www.ms-starship.com).

Some studies cite a possible link between EDCs and population decline in Chondrichtyes species, but further research is needed to clarify the validity of such a link (Gelsleichter et al. 2005; Storelli et al. 2005). Several international bodies, such as the Organization for Economic Cooperation and Development (OECD) and the European Union (EU), have become serious about regulating anthropogenic pollutants that may affect fish. In addition, large research programs, to assist in the development of new guidelines and regulations, have been initiated.

3 The Impact of Anthropogenic Chemicals on Amphibians

The life histories of amphibians are diverse, with some species experiencing complex changes as they are transformed from organisms that live under water (and breath with gills) to forms that occupy the land (and breath with lungs). This metamorphosis process involves structural and biochemical changes that may lead to increased chemical vulnerability at certain life stages. Vos et al. (2000) believed that, because of their transformation processes, amphibians may be at higher risk from anthropogenic chemical exposure than any other vertebrate group.

When amphibians are in their aquatic stage, they are particularly susceptible to xenobiotic exposures. Chemicals may enter the amphibian body through their soft skin, which easily absorbs water. Usually amphibian eggs are laid in water. Therefore, chemicals may harm amphibians during that critical period of development. Chemicals can also interact with the gill-breathing larvae during their aquatic life stage (Gutleb et al. 1999). Terrestrial forms may be affected by anthropogenic chemicals as well.

Inventory, monitoring and experimental studies have been the primary approaches for documenting and discovering the impact of anthropogenic chemicals on amphibian species.

3.1 Reproductive and Developmental Parameters in Amphibians

Several studies point to the threatening role environmental chemicals play on the endocrine system of amphibians. Any alteration of developmental hormones, particularly the thyroid hormones, may have severe consequences for amphibians such as disruption of metamorphosis.

In the majority of cases, amphibians live their early life stages in water. This characteristic renders them particularly vulnerable to EDCs, because water is the main sink for these compounds. In fact, concerns for the environmental affects of endocrine disruptors originally arose because early studies identified their effects on the developing amphibian embryo and foetus.

Although little is known of the possible ecosystem effects of EDCs, several authors believe that these substances may contribute to changes and declines of amphibian populations (Bridges and Semlitsch 2000; Kloas 2002).

3.2 Vulnerability of Early Life Stages

Metamorphosis may render amphibian species more vulnerable to chemicals or toxins; however, little attention has been given to the effects of EDCs on the amphibian species life-history in the context of the overall population dynamics. Bridges (2000) studied the vulnerability of early life stages of the southern leopard frog (*Rana sphenocephala*). Tadpoles were exposed to the pesticide carbaryl at different times during development and it was observed that exposed individuals experienced significant mortality during the early life stages (egg, embryo and tadpole). Delayed metamorphosis of the tadpoles was also an observed effect (Bridges 2000). Embryos and tadpoles of the northern leopard frog (*Rana pipiens*), green frogs (*Rana clamitans*) and North American bull-frogs (*Rana catesbeiana*) were exposed to the insecticide fenitrothion, and the herbicides triclopyr and hexazinone, under laboratory conditions (Berrill et al. 1994). Results showed that newly hatched tadpoles were sensitive to these pesticides; exposures resulted in either death or paralysis, whilst other life stages of these species were almost unaffected. Similar results were observed in embryos and tadpoles of wood frogs (*Rana sylvatica*), American toad (*Bufo americanus*), and green frog (*R. clamitans*) exposed to the insecticide endosulfan. Tadpoles of all three species were paralyzed and post exposure mortality was high (Berrill et al. 1998).

Berrill et al. (1993) further studied embryos and larvae of the wood frog (*R. sylvatica*), northern leopard frog (*Rana pipiens*), green frog (*R. clamitans*), American toad (*B. americanus*) and spotted salamander (*Ambystoma maculatum*), exposed to low concentrations of the pyrethroid pesticides permethrin and fenvalerate. The influence of pesticides was strongest in tadpoles and resulted in delayed growth; tadpole and salamander larvae were twisted abnormally after exposure. Berrill et al. (1993) reported that early life stages of amphibians are likely to be sensitive to even low-level contamination events. Although several studies have revealed that early life stages of amphibians are the most vulnerable stages to environmental chemicals, it is unknown whether the affects on early life stages also produce population effects.

3.3 The Impact of Pesticides and PCBs

Pesticides cause the most severe effects on the amphibian endocrine system. Davidson (2004) conducted the first study, in which population decline in an amphibian species was linked to historical pesticide applications. Results show the that cholinesterase-inhibiting insecticides (mostly organophosphates and carbamates) stood out as more strongly associated with population declines in amphibians than any other pesticide classes (Davidson 2004). In a previous study, other factors such as climate change, UV-B radiation and habitat alteration were evaluated for causing population declines in amphibians as well. Results suggested that the observed declines were not consistent with the climate change hypothesis; results did demonstrate a strong positive association with elevation, percentage upwind agricultural land use, and local urbanization in Central Valley, California, USA (Davidson et al. 2001).

When evaluating endocrine disruption in amphibians, one must remember that many pesticides persist in the environment for long periods, although usually in low concentrations. Storrs and Kiesecker (2004) investigated possible long term (30 d) exposure effects of atrazine on amphibians. Tadpoles of four species of frogs (*Pseudacris crucifer, B. americanus, R. clamitans, and R. sylvatica*) were exposed at early and late developmental stages to low concentrations of a commercial formulation of atrazine (3, 30 ppb or 100 ppb). In all experiments, it was remarkable that survival was significantly lower in individuals exposed to 3 ppm rather than those exposed to 30 ppm or 100 ppm, except for the late stages of the American toad and wood frog tadpoles. Such survival patterns highlight the importance of investigating the impacts of contaminants at realistic exposure levels, and at various developmental stages. This may be particularly important for compounds that produce greater mortality at lower doses than higher ones, a feature characteristic of a number of endocrine disruptors (Storrs and Kiesecker 2004).

Endocrine regulation during metamorphosis comprises several developmental hormones that may be affected by EDCs. In particular, alterations on the thyroid system may result in enhanced or retarded metamorphosis, which may then affect population levels. Kloas (2002), however, pointed out that there is insufficient data to discern whether or not metamorphosis is especially sensitive to the effects of contaminants with endocrine disrupting properties.

Metamorphosis occurs almost universally in all amphibian species. Bridges (2000) reported that any delay or alteration in metamorphosis may impact demographic processes of the population, potentially leading to declines or local extinction.

Theodorakis et al. (2006) studied adult male and female cricket frogs (*Acris crepitans*), in perchlorate-contaminated streams in central Texas, USA, to assess possible endocrine disruption effects on the thyroid system. There was no evidence of colloid depletion or hyperplasia in frogs from any of the sites, although frogs from two sites with the greatest mean water perchlorate concentrations exhibited significantly greater follicle cell hypertrophy. Furthermore, there was a significant positive correlation between follicle cell height and mean water perchlorate concentrations for frogs collected from all sites.

In addition to the thyroid system, the estrogen and androgen systems in amphibians have also been well characterized with regards to their roles in normal development. It is known that the function of estrogens and androgens are subject to perturbation by endocrine-disrupting chemicals, with potential sequelae leading to irreversible consequences for exposed organisms. However, it is known that transient hormone exposure in the adults is reversible. Notwithstanding, little experimental information is available to aid in characterizing the risk of endocrine disrupters on these systems (Bigsby et al. 1999).

It is known that PCBs and DDT can have profound impacts on the estrogen and androgen system in amphibians. Reeder et al. (2005) found that the percentage of intersex in exposed cricket frogs (*A. crepitans*) continually increased, with increasing use of PCBs and DDT in manufacturing and agricultural processes in Illinois, USA. Intersex was highest in heavily industrialized and urbanized northeastern portions of Illinois, and declined with distance from the industrialized areas. It was suggested that these chemicals contributed to the decline of cricket frogs in Illinois.

A study by Mikkelsen and Jenssen (2006) showed that, in adult male European common frogs (*R. temporaria),* PCBs affected the sex hormone homeostasis after the animals were aroused from hibernation. Although no dose-dependent effects were detected, it was assumed that different physiological phases in frogs may be affected by PCBs throughout the yr.

Russell et al. (1995, 1997) linked the decline of different frog species at Point Pelee National Park, Canada, to the heavy use of DDT, until 1967, in this area. The authors surveyed a number of parks and wildlife reserves along the north shore of Lake Erie and found a relationship between the rate of local extinctions of amphibians and the degree of site contamination with chlorinated pesticides. Bridges and Semlitsch (2000) further proposed that chemical contamination, at lethal or sublethal levels, can alter natural regulatory processes such as juvenile recruitment in amphibian populations, and should be considered as a contributing cause of decline in amphibian populations. Berrill et al. (1998) determined that the juvenile aquatic stages of amphibians are sensitive to pesticides, and such pesticides may result in altered swimming performance. The general activity and swimming performance (i.e., sprint speed and distance) of the plains leopard frog (*Rana blairi*) was studied, after acute exposure to carbaryl. Carbaryl greatly affected swimming performance and activity of tadpoles, suggesting that exposure to this carbamate may result in increased predation rates; because activity of this species is closely associated with feeding, carbaryl exposure may result in delayed growth, failure to emerge before pond drying, or result in an indirect reduction in adult fitness. Acute exposure to sublethal toxicants, such as carbaryl, may not only affect immediate survival of tadpoles, but may also affect life history functions and generate changes at the local population level (Bridges 1997).

A prominent pesticide, which occurs at high concentrations in water, and is known to have endocrine disrupting properties, is the herbicide atrazine. Despite its ban in numerous countries, atrazine is still one of the major surface water contaminants in the USA and, to a lesser extent, in Europe. Freeman et al. (2005)

showed that atrazine, at concentrations as low as 100 ppb, increased the time of metamorphosis in the African clawed frog (*Xenopus laevis*) tadpoles. Hayes et al. (2002) examined the effects of atrazine on sexual development in *X. laevis* and found that atrazine induced hermaphroditism and demasculinized the larynges of exposed males. The plasma testosterone levels in sexually mature males were low from a possible conversion of testosterone to estrogen, induced by atrazine. The atrazine levels investigated in this study (0.01–200 ppb) constituted exposures known to exist in nature. Therefore, other amphibian species exposed to atrazine in the wild could be at risk of impaired sexual development, and if true, atrazine may be linked to the global amphibian population decline. Rohr and Palmer (2005) found that streamside salamanders (*Ambystoma barbouri*) exposed to 40 μg/L of atrazine, showed greater activity, fewer water-conserving behaviors and accelerated water loss. Even 4 and 8 mon after termination of exposure, animals were still at a higher risk of desiccation; no recovery from atrazine exposure was detected.

When assessing possible threats to the existence of amphibian populations, it is unfortunate that long-term studies are rare. Such studies will eventually be necessary to understand the long-term effects of chemicals and their possible population effects. Most pesticide effects studies on amphibians are limited to acute or very short-term (4 d) tests conducted under highly artificial conditions. Such studies hardly provide realistic measures of potential field effects. Slightly longer (10–16 d) exposure periods, for example, to the pesticide carbaryl, resulted in a 10–60% higher mortality in gray treefrog (*H. versicolor*) tadpoles. In the presence of predatory stress, the pesticide effect became even more severe. The negative affects to amphibians, in nature, of the pesticide carbaryl may be widespread (Relyea and Mills 2001).

Studies were conducted to test the suspicion that combining the effects of predatory stress and pesticides may produce stronger effects on amphibian populations. Relyea (2004) focused on the effects of malathion, a common insecticide, on tadpoles. Six frog species (*R. sylvatica, R. pipiens, R. clamitans, R. catesbeiana, B. americanus, and H. versicolor*) were studied and malathion was determined to be toxic to all species. The combination of the insecticide plus predatory stress resulted in higher mortality in one of the tested species. These results tended to support the idea that combining exposure to the insecticide carbaryl and predatory stress was synergistic. It was speculated that such synergy may occur with many carbamate and organophosphate insecticides. Relyea (2004) suggested that a combination of pesticide exposure and predator stress may influence amphibian species in a way that alters population dynamics.

Another study addressed the effects of carbaryl on amphibians subject to natural stresses (competition and predation); this study focused on tadpoles of three species: woodhouse's toad (*Bufo woodhousii*), gray treefrog (*H. versicolor*) and green frog (*R. clamitans*). It was observed that carbaryl affected toads and treefrogs in a way that larval survival was reduced. However the effects of carbaryl varied with predator, environment and initial larval density in all species, which, interestingly, resulted in indirect, 'beneficial' effects on amphibians. On the basis of this synecological study, Boone and Semlitsch (2001) stated that interactions of carbaryl with predators may result in the elimination of zooplankton populations that compete with tadpoles for food resources. These results indicate that differences in biotic conditions influence the impact of carbaryl, and that even low concentrations induce changes that may alter community dynamics in ways not predicted from single-factor, laboratory-based studies (Boone and Semlitsch 2001).

Pesticide exposure also influences the behavior of predators. In a laboratory experiment, it was observed that predation of southern leopard frog tadpoles (*Rana sphenocephala*) by adult red-spotted newts (*Notophthalmus viridescens*) was highly dependent on carbaryl concentrations. After exposure to carbaryl for 1 hr, newts consumed half as many tadpoles as nonexposed newts. Carbaryl either affected newt activity in ways that reduced time spent searching for prey, or it may have altered the speed and coordination necessary to capture tadpoles (Bridges 1999).

Pesticides play an important role in the discussion of amphibian population decline. The various ways they influence or impact amphibians are well observed and the majority of authors believe, or have speculated, that pesticides have the potential to cause population declines in amphibian species.

3.4 The Impact of Fertilizers

The high amounts of fertilizers used, especially nitrate that may move to surface and groundwater, is a global problem. Nitrate is known as an important environmental toxicant and it has been shown to impact amphibians in several ways (Guillette and Edwards 2005).

Four tadpole species (*B. americanus, Pseudacris triseriata, R. pipiens, and R. clamitans*) were exposed to ammonium nitrate fertilizer in water. All four species showed toxic effects such as reduced activity, weight loss and physical abnormalities, when exposed to ammonium nitrate at concentrations commonly exceeded in agricultural areas globally (Hecnar 1995).

Marco et al. (1999) reported effects of nitrate and nitrite solutions on newly hatched larvae of five species of amphibians (*Rana pretiosa, Rana aurora, Bufo boreas, Hyla regilla, and Ambystoma gracile*). After exposing the larvae to nitrate or nitrite ions in water, some had reduced feeding activity, swam less vigorously, showed disequilibrium and paralysis, suffered abnormalities and edemas, and eventually died. Even at nitrate concentrations believed to be non lethal (U.S. EPA-recommended limit for warm-water fishes; 5 mg $N\text{-}NO_2^-/L$), and at the recommended limits of nitrite's concentration in drinking water (1 mg $N\text{-}NO_2^-/L$), highly toxic effects were seen in the larvae of all species (Marco et al. 1999).

Although possible nitrate effects on human health are well studied and limits for fertilizers in drinking water are known, their effects on amphibians have received little attention. Although unproved at present, fertilizers may play a substantial role in the apparent global amphibian decline (Hecnar 1995).

3.5 Deformities in Amphibians

In recent yr, large numbers of deformed frogs have been observed throughout North America (Ankley and Giesy 1998; Schmidt 1997). Observed malformations include missing or supernumerary limbs, bony limblike projections, digit and musculature malformations and eye and central nervous system abnormalities (Ankley and Giesy 1998). The most affected species appear to be ranids (*R. pipiens, R. clamitans,* and *Rana septentironalis*). However, it is not yet known whether the instances of amphibian population declines are linked to the observed deformities. A study by Gutleb et al. (1999) reported that a PCB congener caused dose-related malformations (edema, lack of gut coiling, malformed eyes and tails) in embryos of the African clawed frog (*X. laevis*) and that retinoid concentrations were significantly altered in PCB-dosed embryos.

Further research is needed to elucidate whether or not environmental contaminants are responsible for the observed malformations in amphibians and, if so, whether or not these effects are mediated through endocrine disrupting mechanisms.

3.6 Impact on Amphibians: Conclusion

It cannot be denied that amphibian populations are declining dramatically in many areas of the world (Pechmann et al. 1991; Pechmann and Wilbur 1994). A study by Houlihan et al. (2000) examined the high rate of amphibian decline globally. He catalogs large declines beginning in the late 1950s, and continuing until the early 1960s, followed by a falling decline rate up to the present. Their studies confirmed suspicions that amphibian populations are declining, albeit with geographical and temporal variability. Even if the cause or causes of the decline are unknown, many believe they result from man-made alterations in the environment (Houlihan et al. 2000). Much effort was undertaken in the last few yr to explain why amphibian populations are declining, in both pristine and polluted habitats worldwide (Vos et al. 2000).

Although loss of habitat is known to affect amphibian population decline, recent research has focused on the effects of environmental contaminants, UV-B irradiation, emerging diseases, the introduction of alien species, direct exploitation, and climate change (Beebee and Griffiths 2005).

Pesticides may be a possible cause of, or contributor to, amphibian population decline, but pesticide research with amphibians has focused mainly on single organism tests. Nevertheless, pesticides have the potential to alter developmental processes, affect reproduction, and are known to be particularly harmful to early life stages. In an attempt to explain the decline, some researchers have tried to extrapolate observed effects of pesticides gleaned from small laboratory studies to affects on whole populations. However, to confirm the merits of such an extrapolation, long term studies with pesticides on population dynamics are necessary. Because of a dearth of relevant data, distinguishing between natural fluctuations in population size and structure, and anthropogenic-induced declines is challenging (Pechmann et al. 1991).

Many believe that the recently reported deformities in amphibians result from exposure to environmental chemicals. However, it is not known if chemical exposure has contributed to the global decline in amphibian populations, or not.

There is a hypothesis that more than one factor is responsible for the decline in amphibian populations. Beebee and Griffiths (2005) and Vos et al. (2000) suggest that an interaction between abiotic and biotic factors have the potential to cause the declines. However, different species and different populations of the same species may react in different ways to the same environmental insult. Species with declining populations are often found in environments that are physiographically similar to those where the same species is thriving.

4 The Impact of Anthropogenic Chemicals on Reptiles

The diversity of reptiles, including Crocodilia (crocodiles, caimans and alligators), Sphenodontia (tuataras), Squamata (lizards, snakes) and Testudines (turtles), is enormous; they are found on every continent except Antarctica.

Reptiles, known as ectotherms, are strongly dependent on their natural habitat, and any environmental disturbance such as habitat destruction can have profound effects on the survival of affected individuals. The potential effects of anthropogenic chemicals on reptiles are dependent on the nature of their reproductive and developmental strategies, which are highly diverse (Lamb et al. 1995; Palmer et al. 1997). For example, most species are oviparous, but ovoviviparity or viviparity also occurs (Palmer et al. 1997); even among the oviparous species, the female reproductive tract exhibits great anatomical variation (Palmer and Guillette 1988, 1990, 1992).

Most oviparous reptiles bury their eggs and, in so doing, create a possible pathway for chemicals to impact early life stages. Although the eggs are surrounded by an eggshell, dissolved compounds may enter the egg. Among viviparous species, the potential for biomagnification in the young is high, because the adults nourish their young through various forms of placenta.

Many reptiles such as turtles, crocodiles or large snakes may live for decades. The life expectancy is often more that 30 yr in such animals, and up to 150 yr for some species. The dietary requirement for reptiles is dependent on the species. Some are herbivores and some are carnivores, with many carnivores existing at, or near the top of the food web (Bowler 1977; Congdon et al. 1983; Gibbons and Semlitsch 1982). Top feeding reptilian carnivores are highly vulnerable to environmental toxins, particularly because of the potential they have to bioaccumulate and biomagnify consumed chemicals (Cobb and Wood 1997; Hall and Henry 1992; Olafsson et al. 1983).

4.1 The Impact of EDCs on Gender Determination

Reptiles have long been used as good bioindicators of environmental contaminants. Reptile species have increasingly become more interesting as targets for studying the mechanisms by which endocrine disrupters act; this is because different species have varying gender determination which makes them good models for studying the impact of EDCs.

The mechanisms that determine gender in reptile species are well understood. In some species, hormones influence specific structures that will ultimately differentiate between the sexes after the formation of the gonads. Alternatively, it is known that many egg-laying species (e.g., crocodiles, turtles and lizards) do not exhibit genotypic sex determination. The sex of the offspring is dependent on the incubating eggs response to temperature. This phenomenon is called temperature-dependent sex determination (TSD). The gonadal sex is not ultimately set by the genetic composition inherited at fertilization, but depends on the temperature-dependent pattern of activation of those genes, which encode for steroidogenic enzymes and hormone receptors during embryonic development (Lance 1994). The pattern of TSD among reptile species is highly divergent, which complicates attempts to understand the possible alterations induced by EDCs on the sex determination process (Wibbels et al. 1998).

Crews et al. (1995) studied reverse gonadal sex in turtles exposed to steroid hormones. These hormones overrode the effects of temperature, and led to altered sex determination at a temperature that otherwise would have produced males. Developing fence lizard (*Lacerta agilis*) embryos exposed to an estrogenic chemical under laboratory conditions showed similar results. Eggs injected with the estrogenic chemical 17α-ethinylestradiol led to a feminization of males and prevented development of embryonic secondary sex characteristics (Talent et al. 2002).

It is known that some environmental chemicals, particularly organochlorines, mimic the effects of natural hormones. This has been well studied in other non-reptile species such as fish (Kime et al. 1999; Tyler and Routledge 1998) and may have further implications on wildlife population dynamics.

It is also known that some pesticides have endocrine disrupting properties and may profoundly affect expected sex outcomes in reptiles. Red-eared slider turtles (*Trachemys scripta elegans*) exposed to one of three pesticides (chlordane, trans-nonachlor or DDE (dichlorodiphenyl dichlorethylene)) during embryogenesis produced altered sex determination and sexual development (Willingham 2001). It was also observed that all three compounds produced certain population-wide effects (changes in hatchling body mass), when compared to controls. Willingham (2001) suggested that these results point to a role for pesticides in endocrine disruption that extends beyond sex determination and sexual development.

Portelli et al. (1999) studied eggs of the common snapping turtle (*Chelydra serpentina serpentina*), a species with temperature-dependent sex determination, during embryonic development in the Great Lakes, USA. This species was exposed to the pesticide metabolite DDE at doses (0.52–65 μg/5 μl ethanol) selected to simulate concentrations found in the Great Lakes. It was expected that DDE has profound impact on the sex determination in the common snapping turtle. Results of this study, however, revealed that DDE did not affect sex determination at the exposure levels used. The results further indicate that DDE, at levels found in the environment in the Great Lakes, does not cause feminization of snapping turtles during embryonic development.

4.2 The Impact of EDCs on Reptile Populations

Many scientists believe that EDCs are responsible for, or contribute to, the observed population declines of reptiles. A prominent case of a possible link between EDCs and population decline comes from the American alligators (*Alligator mississippiens*) in Lake Apopka, USA. Lake Apopka is a hypertrophic lake in Florida, USA with a 50-yr history of contamination from agricultural and municipal sources. In 1980, a stream that feeds Lake Apopka was contaminated with high concentrations of dicofol and other DDT congeners after a chemical spill. In the following yr (1980–1984), the population of American alligators (*Alligator mississippiens*) declined by 90% (Guillette et al. 1994). The decline was attributed to the EDCs DDT and DDE, and the observed developmental abnormalities (altered gonadal steroidogenesis, abnormal gonadal morphology and changes in sex steroid concentrations in males and females) in juvenile alligators confirmed the suspicion.

In 1984 it was observed that Lake Apopka alligator tissues contained concentrations of DDE, dieldrin, endrin, mirex, oxychlordane, DDT and PCBs (Guillette et al. 1999). These compounds were said to be responsible for the observed decline in clutch viability, effects which linger today. Furthermore, the effect of the broad-spectrum insecticide toxaphene (found in relatively high concentration in Lake Apopka alligator egg yolk) on alligator gonadal development were tested. Toxaphene failed to affect sexual differentiation and did not induce developmental abnormalities (Milnes et al. 2004). These results suggest that, to better evaluate consequences of environmental contamination, more attention must be focused on testing the effects of chemical mixture exposures on embryonic development in alligators (Milnes et al. 2004).

Studies with other chemicals that focused on different locations and species produced similar results. For example, western pond turtle eggs (*Clemmys marmorata*) from Fern Ridge Reservoir in western Oregon were contaminated with high levels of organochlorine pesticides, PCBs and metals. Contaminated eggs failed to hatch. It was suggested that these contaminates may account for the decline of the western pond turtle population in this area (Henny et al. 2003).

Wu et al. (2000) studied the impact of organochlorine compounds (lindane, aldrin, methoxychlor, heptachlor epoxide, DDT) on eggs of Morelet's crocodile (*Crocodylus moreletii*) from Gold Button and New River lagoons in northern Belize. Based on the results of 24 analyzed egg samples, it was proposed that organochlorine-exposed crocodiles from both lagoons may suffer threats to health that could impair population dynamics of crocodiles in Central America. Crain and Guillette (1998) claimed that contaminant-induced endocrine alteration in reptile embryos may lead to impaired reproduction, which in turn can affect population dynamics. Henny et al. (2003) studied eggs of the common snapping turtle (*C. serpentina serpentina*) near the Great Lakes–St. Lawrence River basin, USA. Although the eggs were contaminated with organochlorine pesticides and PCBs that are known to produce effects on sex differentiation and reproductive endocrine function, other reasons were given for the observed effects.

It is commonly accepted that reptiles and other natural biota may be simultaneously influenced by more than one chemical. Moreover, it is generally accepted that different factors together can have profound effects on population dynamics. Willingham (2005), for example, used embryos of the red-eared slider turtle (*Trachemys scripta elegans*) to show the possible combined effects of increased temperature and the herbicide atrazine on sex ratio. He observed that increased temperature or atrazine alone did not affect sex ratio. However, if the two factors interacted, the female fraction significantly increased. This result proved that, at least in some cases, a combination of two or more factors is necessary to exert an observed effect.

4.3 Developmental Abnormalities in Reptiles

The common snapping turtle (*C. serpentinia serpentinia*) inhabits large regions of the Great Lakes–St. Lawrence River basin, USA. In recent yr, an increased level of developmental abnormalities was detected in this species (Bishop et al. 1998). Bishop et al. (1991) reported that the large number of unhatched snapping turtle embryos and hatchling deformities observed in those animals could be linked to chlorinated hydrocarbon exposure in the river basin. PCBs, PCDDs (polychlorodibenzodioxins) and PCDFs (polychlorodibenzofurans) were particularly accused for the observed abnormalities, which included absent or altered tails, carapace anomalies (missing or extra scutes), unresorbed yolk sacs and fore and hind limb deformities (Bishop et al. 1998). The observed abnormalities may be illustrative of additional EDC effects on reptiles. The induction of these abnormalities may also contribute to the declining population numbers of common snapping turtles in the Great Lakes–St. Lawrence River basin (Bishop et al. 1998).

4.4 Impact on Reptiles: Conclusion

Many studies have emphasized the effects of anthropogenic chemicals, especially EDCs, on the physiology and reproduction of reptiles. However, most of this work has been restricted to laboratory studies. Much less work has been conducted to quantify the effects of toxic chemical exposures on reptiles in the wild.

Endocrine disrupting chemicals manifest their effects on the endocrine systems of the animals exposed to them. Because endocrine systems are highly divergent across the reptilian class, predicting the physiological responses of xenobiotic exposures on reptilian species is very difficult. Although some reptile populations have been affected by EDCs, it is unlikely that all reptilian species are equally sensitive to the effects of EDCs. However, among the various reptile species studied, results indicate that adults and embryos are currently experiencing toxic effects and, in some species and locations, there is evidence that population declines are caused or triggered by environmental chemical exposure (Bishop and Gendron 1998; Bishop et al. 1998).

5 The Impact of Anthropogenic Chemicals on Birds

Birds (Aves) include more than 9,000 species. Bird activities such as courtship, breeding, migration, etc., require high energy expenditure and, because birds also have high metabolic rates, large amounts of food are necessary to make survival possible. The diet of birds is strongly dependent on the species; insectivores, carnivores, piscivores, herbivores, and also fruit eaters, are known. At periods during their seasonal cycles, many bird species experience starvation; for example while breeding or when migrating to different habitats. Birds respond to starvation by mobilizing stored lipids. Any lipophilic chemicals stored in bird adipose tissue is then easily released to the systemic circulation, and may harm the organism.

It was observed that birds are particularly vulnerable to environmental chemicals (e.g., organochlorines) during early life. PCBs and various pesticides also pose major threats to some bird species. Waterfowl are particularly susceptible to accumulation of persistent organic pollutants (POPs) that are known to constitute a major hazard for birds (Giesy et al. 1994b). Persistent and bio-accumulative organic compounds were found in high concentrations in waterfowl. Such compounds exert severe effects on reproduction and, simultaneously may be responsible for deformities and mortality (Giesy et al. 1994a, b; Gilbertson 1983). Oaks et al. (2004) reported a decrease exceeding 95% of the oriental white-backed vulture (*Gyps bengalensis*) population in the Indian subcontinent, since the 1990s. The authors observed that birds exposed to diclofenac suffered from renal failure and visceral gout. The authors concluded that residues of veterinary diclofenac are responsible for the observed vulture population decline.

It was reported by several bird watch organizations that many bird populations, worldwide, are declining (Worldwatch Institute 2003; http://www.worldwatch.org). The reasons for the observed decline are not fully understood. However, factors such as EDCs, habitat loss, predation by non-native species, oil spills and pesticide use, industrial pollution and climate change are generally accepted to cause or contribute to bird population declines.

5.1 Reproductive Parameters in Birds: Behavior and Sexual Differentiation

DDT, PCBs and mixtures of other organochlorines have been identified as EDCs. These chemicals have the potential to affect reproduction of bird populations in many areas worldwide, and they have been linked to global bird population declines (Peakall 1986, 1988). Bird embryos are most endangered from exposure to EDCs. Exposure during early life stages can result in mortality, failure of chicks to thrive and impaired differentiation of the reproductive and nervous systems through mechanisms of hormonal mimicking of estrogens (Fry 1995). Effects of EDCs on adult birds include acute mortality, sublethal stress, reduced fertility, suppression of egg formation, eggshell thinning and impaired incubation and chick rearing behaviors (Fry 1995).

In birds, the gonads and bird behavior are both affected by the differentiating hormone estrogen. As mentioned, EDCs have the potential to mimic hormones such as estrogens or androgens and, therefore, can profoundly affect bird behavior and sexual differentiation. Adkins-Regan et al. (1994) suggested that sexual differentiation in birds is sensitive to estrogens and androgens. They further believe that any hormonal disturbance can produce unpredictable effects on reproductive behavior in both sexes. Adkins (1979) studied male Japanese quail embryos (*Coturnix coturnix japonica*) treated with estrogens before d 12 of the 18-d incubation period, and discovered that the quail suffered from dramatic sex-reversing effects. In contrast, the estradiol-induced masculinization in female zebra finch (*Taeniopygia guttata*) is produced only after hatching (Adkins-Regan et al. 1994).

Japanese quail are a precocial species (birds that are in an advanced state of development at hatching), whereas, zebra finches are typically altricial songbirds (birds in an early state of development at hatching). When studying effects of xenobiotic exposures, one must consider the implications of birds being precocial or altricial species; they develop similarly but hatch at different times of the overall developmental sequence (Adkins-Regan et al. 1994).

There are several studies that address behavioral alterations of adult birds exposed to chemicals. Barron et al. (1995) reported that sublethal levels of PCB result in reduced parental attentiveness and abnormal reproductive behavior in free living birds. Fox et al. (1978) observed abnormal parental behavior (failure to sit on eggs or to defend nests) in herring gulls (*Larus argentatus*) exposed to organochlorines. The high levels of endocrine disrupting chemicals found in these birds were said to be responsible for the observed behavioral abnormalities.

The consumption of a mixture of DDE and PCBs led to a reduction or delay in behaviorally induced increase of sex hormones in adult ring doves (*Streptopelia risoria*). McArthur et al. (1983) observed that exposed females showed altered courtship behavior, did not respond to male courtship, and spent less time in feeding their young. Organochlorines (DDE, PCBs) were believed to be responsible for the altered hormone levels and unusual reproductive behavior seen in these ring doves.

Female adult ring necked doves (*Streptopelia capicolas*) showed depressed courtship behaviors (Tori and Peterle 1983) after exposure to PCBs; this resulted in reduced reproductive success and aberrant breeding (Peakall and Peakall 1973). A study by Bennett et al. (1991) showed that the insecticide parathion can affect bird populations in ways that lead to altered incubation behavior and reduced reproductive success. Furthermore, females of lesser scaup (*Aythya affinis*) (while migrating or over wintering) experienced lower survival, altered reproduction and reduced-courtship behaviors, after exposure to dietary contaminants existing in exotic bivalves (Fox et al. 2005). In this study, eggs and nestling females of lesser scaup were analyzed for environmental contaminants. It was determined that zebra mussels (*Dreissena polymorpha*) and Asian clams (*Potamocorbula amurensis*), predominant prey species of this bird, contained high concentrations of selenium. The concentration of selenium consumed may have affected courtship behavior, caused sublethal effects, and possibly mortality, when eaten by scaups for some time. Furthermore, it was proposed that the continental decline of these birds in boreal forests of Canada and Alaska may be linked to the high consumption of mussels contaminated with EDC (Fox et al. 2005).

We conclude from the foregoing, that chemicals, particularly EDCs, are currently affecting behavior and sexual differentiation in some bird species.

5.2 The Impact on Reproductive Organs

In addition to altering behavior in birds (Fox 1992; Gilman et al. 1979), many authors have also observed EDC effects on reproductive organs. Fox (1992) studied male herring gull (*L. argentatus*) embryos from Scotch Bonnet Island, Ontario, Canada and found that about 57% suffered from testicular feminization. Eggs of this species contained high levels of dioxins and PCBs (Fox 1992; Gilman et al. 1979); it was hypothesized that these contaminants may have caused the impaired gonadal development. In contrast, the reason behind a high rate of abnormality in testes of terns (*Sterna forsteri*) was not identified (Nisbet et al. 1996). Fry et al. (1987) studied adult female herring gulls (*L. argentatus*) collected from Tacoma, Washington, adjacent to the Commencement Bay, Puget Sound (a PCB- and heavy metal-contaminated superfund site), in 1984. The right oviducts of these gulls (which are reduced during normal avian ontogeny) were found to be enlarged and to persist longer than normal. The length of the right oviduct was correlated with the level of estimated chemical contamination (Fry et al. 1987). The relevance of this observation is unclear, because all birds were successfully breeding (Boss and Witschi 1947).

Feminization of gonads of male embryos and persistence of right oviducts in female embryos were observed in experimental studies of western (*Larus occidentalis*) and California (*Larus californicus*) gull eggs injected with hormones and other substances. The hormones tested included estradiol (Fry and Toone 1981) and diethylstilbestrol (DES), a synthetic estrogen (Boss and Witschi 1947). Among the tested environmental contaminants were methoxychlor and DDT (Fry and Toone 1981). Because concentrations of DDT (2–100 ppm) found in the eggs of wild gulls caused effects consistent with those induced by estradiol and DES (Fry and Toone 1981), it was suggested that DDT or other estrogenic contaminants could be responsible for the effects observed in the wild. Whether or not the observed effects impaired the reproductive success of adult birds was unclear (Fry and Toone 1981; Fry et al. 1987). In a different study, altered gonadal development and ovotesis formation in male embryos of western (*Larus occidentalis*) and California (*Larus californicus*) gulls was detected when exposed to 17β-estradiol, DDT and environmental contaminants (NRC 1999). In summary, EDCs probably cause alterations of reproductive organs in birds, but it is, as yet, not known whether these impacts have population-wide consequences.

5.3 Great Lakes Embryo Mortality, Edema, and Deformities Syndrome

Fish-eating birds (herring gulls (*L. argentatus*), common terns (*S. forsteri*) and double-crested cormorants (*Phalacrocorax auritus*)) that live in the Great Lakes basin, in North America suffer from a syndrome called GLEMEDS (Great Lakes Embryo Mortality, Edema, and Deformities Syndrome) (Gilbertson and Fox 1977; Gilbertson et al. 1991). GLEMEDS involves developmental abnormalities, including bill deformities, club feet, missing eyes, defective feathering, liver enlargement, liver necrosis (Fox 1991; Gilbertson et al. 1991; Ludwig et al. 1993) and other abnormalities that are of ectodermal origin (Rogan et al. 1988). The hypothesis has been advanced that GLEMEDS, in colonial fish-eating birds, resembles chick-edema disease of poultry and has been caused by exposure to chick-edema active compounds (mainly dioxins) that have a common mode of action through the cytochrome P450 system (Gilbertson et al. 1991). In the Gilbertson et al. (1991) study, the authors observed that, with declining concentrations of DDT, PCBs and PCDDs/PCDFs in the Great Lakes, the populations of herring gulls, double-crested cormorants and other fish-eating birds increased. The rate of reproductive failure and the symptoms of GLEMEDS have also decreased with time (Grasman et al. 1998). However, in some regions of the Great Lakes, symptoms of GLEMEDS persist, especially in fish-eating water birds (Fox 1991, 1993). The strong temporal association of GLEMEDS with the presence of PCBs and dioxins implies a causal relationship between environmental pollution and the syndrome.

5.4 Altered Sex Skew

Gull populations from the USA and Canada displayed an altered sex ratio (overabundance of females) and female-female pairings, in some colonies. This phenomenon is detected by documenting the number of nests that contain five or more eggs (supernormal clutch). A single female gull typically lays one to three eggs (Conover et al. 1979). The supernormal clutches result from polygynous trios of two females and one male (Conover and Hunt 1984a).

The most dramatic and well-documented example of altered sex skew occurred in the western gull population on Santa Barbara Island in California from 1968 to 1978 (Hunt et al. 1980). Female–female pairings reached 15% of all pairing individuals. Supernormal clutches were also observed in herring gulls (*L. argentatus*) that inhabit the northeastern portion of Lake Michigan, USA (Fitch and Shugart 1983; Shugart 1980).

The California and the Great Lakes gull populations were both exposed to great levels of organochlorine contamination, including DDT, during the 1950s–1970s (Fry and Toone 1981). The sex skew favored females in both populations. Although it was reasonably assumed that DDT and other organochlorines are responsible for the observed effects, the causal link was not established. In contrast, a study showing an incidence of supernormal clutchs in Caspian terns (*Hydroprogne caspia*), ring-billed (*Larus delawarensis*) and California gulls (*Larus. californicus*) is known to have occurred before the DDT era, and their frequency has not changed over time (Conover and Hunt 1984a). Consequently, other factors may be responsible for the observed female-female pairings, as well.

Notwithstanding, as DDT levels have declined in the environment, it was discovered that the incidence of supernormal clutches has decreased significantly for many species of terns throughout the USA (Conover and Hunt 1984b). Hence, it is probable that DDT induces supernormal clutches, even though it is not the only reason behind this phenomenon. A shortage of males during the breeding season is one possible reason for the observed abnormalities. Indeed, Conover and Hunt (1984a) indicate that female-female pairings allow females to breed when they are unable to obtain a male partner. Sex skew toward females in western and herring gulls could result from a differential mortality between males and females. It is also possible that male gulls may be more susceptible to, or more rapidly accumulate chemicals, because they are higher up the food chain; if true, this may also explain the higher male mortality. However, such speculations are not well documented (Pierotti 1981).

The underlying cause for the observed alterations in sex ratio and female-female pairings in some bird populations is not clear. Although the results indicate that endocrine disruption may play an important role, other factors such as a shortage in male birds or other reasons can not be ruled out.

5.5 Eggshell Thinning

Eggshell thinning caused by organochlorine pesticides, such as DDT and its degradation product DDE, is known to be species-dependent. It is known that the species exposed to the largest amount of DDT are unfortunately those that are the most sensitive to the insecticide.

A diet of only a few parts per million of DDT will cause 20% eggshell thinning (the degree of thinning that causes eggshell breakage and thus reproductive failure) in raptors and in some fish-eating bird species, such as the brown pelican (*Pelecanus occidentalis*). The pelican population along the Pacific, Atlantic and Gulf Coasts of the US, dramatically decreased between 1960 and 1969 as a result of cracked or broken eggs and other adverse reproductive effects (Elliott et al. 1988; Struger and Weseloh 1985; Struger et al. 1985). Other species that breed in North America, such as the white-tailed eagle (*Haliaeetus leucocephalus*), the osprey (*Pandion haliaetus*) and the cormorant (*P. auritus*) also suffered from high egg breakage, population decline, and near total reproductive failure up until 1972 (Weseloh et al. 1983). Many of the species in which eggshell thinning was observed experienced an increase in population size, after DDT was banned, suggesting that this insecticide was responsible for the observed effects (Bignert et al. 1994; Ludwig 1984; Price and Weseloh 1986; Weseloh and Ewins 1994).

Studies on Canadian and Russian peregrine falcon populations (*Falco peregrinus*) (Johnstone et al. 1996) and some sparrow hawk (*Accipiter nisus*) populations in North America (Fent 2003) reveal that, even today, eggshell thinning is a problem as a result of high DDT content in eggs. Other adverse effects, such as localized impairment of reproductive performance (Tillitt et al. 1992) and anatomical defects (Giesy et al. 1994b), have persisted in some populations and may be another manifestation of lingering environmental residues of DDT and its metabolites. In contrast to the foregoing, some bird species such as gull, terns, and ducks are only moderately sensitive to DDE (Barrett et al. 1997). Some species (e.g., quail and chicken) are nearly insensitive to DDE-induced eggshell thinning. It was impossible to experimentally achieve more than a few percent thinning, even at the highest dosage, without causing mortality to these species (Barrett et al. 1997).

In summary, it is accepted that DDE-induced eggshell thinning has been responsible for the decline of many raptorial bird species. Further, population declines may occur in some sensitive species, while others are almost unaffected by intake of DDE.

5.6 The Impact of Oil Spills

Oil spills is another factor that may cause wildfowl population decline in some species. After the 1989 'Exxon Valdez' oil spill in Prince William Sound, Alaska, at first estimate, ~300,000 birds were affected by the accident. Of major biological significance, was the death rate among white-tailed eagles (*Haliaeetus albicilla*). Four mon after the accident, about 35,000 birds of this species were found dead (Fent 2003).

A study on harlequin duck (*Histrionicus histrionicus*) populations focused on the status of recovery after the accident. It was observed that the population had not fully recovered 9 yr after the oil spill. This observation contrasted with the conventional paradigm that oil spill effects on bird populations are short-lived

(Esler et al. 2002). The populations densities before and after the oil spill were monitored in other species, as well. These species included: (pigeon guillemot (*Cepphus columba*), black oystercatchers (*Haematopus bachmani*), black-legged kittiwakes (*Rissa tridactyla*) and glaucous-winged gulls (*Larus glaucescens*). The spill had a negative effect on population levels of all species, and population numbers had not recovered to pre-spill levels 9 yr after the oil spill. The failure of a complete recovery may have resulted from the persistence of residual oil remaining in the environment, and reduced abundance of forage fish (Irons et al. 2000).

It was observed that taxa of marine birds that prey on fish have declined in Prince William Sound; however, most taxa that feed on other prey species such as benthic invertebrates, have not declined. Similar effects were also documented, over the past two decades, in the Gulf of Alaska, the Bering Sea and along the California coast; the reason is thought to be linked with changes in forage fish species in the North Pacific Ocean. Many declines appear to be related to changes in forage fish abundance that occurred during a climatic regime shift in the North Pacific Ocean, although some taxa were also affected by the Exxon Valdez oil spill (Agler et al. 1999).

The 2002 'Prestige' oil spill offshore from Galicia, Spain, also led to a mass mortality of sea birds. More that 115,000 birds were found dead. After the accident a reduction in the reproductive success of many species occurred, with the European shag (*Phalacrocorax aristotelis*) being particularly affected (Velando 2005). Most species recovered in the yr following the accident.

In 1999, during the 'Erika' oil spill in Brittany, more than 100,000 birds died, and sea bird populations declined dramatically. Common guillemot (*Uria aalge*) populations were the main ones affected by this accident (Castelege et al. 2004).

In summary, sea bird populations are severely affected by oil spills from tanker accidents. Despite the fact that the main effects are transient, a full recovery of affected populations takes considerable time; in some cases, recovery did not occur, even a decade after the spill.

5.7 Impact on Birds: Conclusion

There is ample evidence that some bird populations are declining and some species will face extinction in coming yr. This evidence also supports the premise that bird populations are affected by environmental chemicals. The majority of studies emphasize the impact of EDCs. EDCs may have profound effects on populations when they induce reproductive system effects or alter behavior of exposed organisms. Dramatic effects such as DDT-(DDE)-induced egg-shell thinning were observed and were linked to the decline of affected populations. However, not all bird species are sensitive to DDT. In fact, common test species are insensitive to DDT and, thus, even if the measurement of eggshell thinning had been included in test protocols for new pesticides, this phenomenon would not then have been discovered.

The main effects of oil spills on bird populations are usually regarded to be transient. In the yr that follow environmental oil spills, bird populations tend to recover, though full recovery may require a decade or longer. Unfortunately, the long lasting effects of oils spills have not been investigated in many species. Although not properly tested, oil may have endocrine disrupting properties on organisms. One factor in their favor, is that the majority of bird species seem to be able to migrate to less polluted habitats after an accident occurs.

Despite the number of studies that have been conducted, the causative agents and underlying mechanisms responsible for the observed declines in wild bird populations are basically unknown. Even low doses of contaminants may act as stressors that, in combination with other stressors, could affect bird populations. However, apart from the observed DDT/DDE-induced eggshell thinning, a link between a distinct chemical and the typical effects encountered, is still missing.

6 The Impact of Anthropogenic Chemicals on Feral Mammals

Mammals comprise a group of roughly 5,000 species and represent one of the most dominant groups of living terrestrial and aquatic vertebrates. Their morphological diversity is enormous and they possess a wide array of anatomical, physiological and behavioral strategies (Vaughan et al. 2000).

Egg laying is a very primitive form of mammalian reproduction, and is only represented by monotremes such as the duckbilled platypus and the echidna (Pough et al. 2004).

In contrast, marsupials, including kangaroos, wombats and opossums, and their embryos are born live, but in an extremely immature state. Essentially, a helpless embryo climbs from the mother's birth canal to the nipples. There it attaches with its mouth, eats and continues to develop, often for week or mon depending on the species. The short gestation time results from having a yolk-type placenta in the mother marsupial (Pough et al. 2004).

The vast majority of mammalian species are placental mammals, the progeny of which are born at a relatively advanced stage and then develop unattached to the mothers' body. Characteristically almost all mammalians feed the young by producing milk (Pough et al. 2004). This characteristic represents a first important pathway for chemicals to enter mammalian juveniles during critical periods of their development.

Mammals, who occupy high trophic levels, can accumulate large amounts of persistent chemicals through consumption of prey, who, themselves have accumulated chemicals through biomagnification (Tanabe et al. 1988). Chemicals such as organochlorines, pesticides, PCBs and other lipid-soluble compounds easily bioaccumulate in lipid-rich tissue or blubber of those animals.

Fish-eating mammals may be particularly vulnerable to environmental contaminants, because they often inhabit contaminated coastal or river estuary areas that are polluted by industry effluent or agricultural runoff. In these areas, contaminant burdens are generally higher than in the open ocean, and these levels are further enhanced through bio-accumulation and bio-magnification processes.

6.1 Endocrine Disruption in Marine Mammals

In mammals, EDCs may act as modulators, inhibitors of hormone metabolism, or as alternate ligands that bind endogenous hormones *in situ*. EDCs may also interfere with signalling subsequent to receptor-ligand binding, because they serve as target organ toxicants or modulators of central nervous system components responsible for neuroendocrine regulation (Barton and Andersen 1998). EDC exposure has resulted in both reproductive and non-reproductive effects in organisms.

Severe population declines in Baltic ringed (*Phoca hispida botnica*) and gray seals (*Halichoerus grypus*) were observed in studies conducted over the last 100 yr (Bergman and Olsson 1985; ICES 1992). During the 1960s and 1970s, evidence was provided for Baltic ringed seals, which indicated organochlorines were affecting female reproductive organs in a way that greatly reduced reproductive success (Bergman et al. 2001). Severe claw malformations, arteriosclerosis, uterine cell tumors, and decreased epidermal thickness in Baltic ringed seals and grey seals were reported (Bergman 1999a, b; Bergman and Olsson 1985), and were attributed to PCB and DDT exposure (Lund 1994). Juvenile grey seals sampled along the Swedish Baltic coast showed high concentrations of PCB and DDT in their tissues. Roos et al. (1998) regarded the organochlorines, which existed in those species, to have potential population dynamics affects.

Kostamo et al. (2002) was able to attribute the decline of a Saimaa ringed seal (*Phoca hispida saimensis*) population, in Finland, to the high levels of organochlorines and Hg in their habitat. The decline of the harbor seal (*Phoca vitulina*) population in the Dutch Wadden Sea may also be linked to organochlorine exposure, particularly PCB exposure (Reijnders 1980). It was further suggested that the population decline was a result of low reproduction success, probably caused by consumption of polluted fish (Reijnders 1986,1990).

Although over hunting and habitat destruction may have been contributing factors for the population decline in these species, it is generally accepted that persistent pollutants, which adversely affect the reproductive performance of females, result in declining seal numbers.

The grounding of the 'Exxon Valdez' oil tanker in Alaska, in 1989, had severe effects on seal populations. It was observed that the number of harbor seals (*P. vitulina*) in eastern and central Prince William Sound has been declining since the accident, with an overall population reduction of 63% through 1997 (Frost et al. 1999). Fair and Becker (2000) suggested that the observed population decline in seal populations resulted from acute and chronic effects of the oil spills, but other environmental contaminants and fishery-induced stress may also have produced chronic effects. The tanker accident also had severe consequences for the abundant sea otter (*Enhydra lutris*) population. Otters were the mammalian species most affected by the tanker accident; 4,000 dead otters were found, even 4 mon after the disaster.

Two studies addressed the recovery of the sea otter population after the tanker spill, but produced contrasting results. One of these studies detected a steady increase in otter populations in the yr after the accident. Between 1990 and 1996 the otter population was reported to be higher than before the accident. A decline in population was only seen outside the area that had received the burden of residual oil, in the northern parts of Prince William Sound (Garshelis and Johnson 2001). Results of the other study, conducted at northern Knight Island, where oil burdens were heavy, reported that sea otter abundance was reduced by a minimum of 50%. Even between 1995 and 1998, 6 and 9 yr, respectively after the spill, the size of this population was reported to be less than before the accident (Dean et al. 2000). A study by Burn and Doroff (2005) discovered a continuing decrease in sea otter abundance along the Alaska Peninsula between 1986 and 2001.

The impact of PCBs and DDT on California sea lions (*Zalophus californiansus*) was addressed in other studies. DeLong et al. (1973) found stillbirths and premature pupping in this species, and linked the observed effects to high PCB and DDE levels in their habitat. Gilmartin et al. (1976) believed the described effects could have resulted from diseases such as leptospirosis and calcivirus infections, which have much the same effect on sea lions as EDCs. In elaborating this idea, EDCs may have affected immune functions, which in turn led to disease outbreaks (Gilmartin et al. 1976). However, more research is needed to clarify and confirm this concept.

Beluga whales (*Delphinapterus leucas*), living in a section of the St. Lawrence River in North America which contained high levels of organochlorine pollutants, showed signs of hermaphrodism. This effect was attributed to PCB/DDT-related hormonal disturbances that occurred during early pregnancy (De Guise et al. 1994). Normal differentiation of male and female organs was disrupted. However, no population-level effects were detected in this study.

Environmental chemicals can produce effects of a non-reproductive nature, and several studies have reported such effects on wildlife mammals. Although more data are needed, some evidence suggests that EDCs can profoundly affect the immune system. A possible example from marine mammals is the serious disease outbreaks that occurred in recent yr among seals, sea lions, and dolphins. The disease outbreak was attributed to possible contaminant-related immune suppression (De Swart et al. 1995). De Swart et al. (1994) observed altered natural killer cell activity and T-lymphocyte function in female harbor seals (*P. vitulina*) from the Wadden Sea, Holland. The diet of those seals contained high amounts of contaminated fish. Harbor seal, Baikal seal (*Phoca sibirica*), striped dolphin (*Stenella coeruleoalba*), and bottlenose dolphin (*Tursiops truncata*), observed by Dietz et al. (1989), showed similar effects. It was suggested that the uptake of contaminates led to immune suppression in those animals. It was further proposed that contaminant-induced immune suppression could have contributed to mass mortalities of marine mammals (Dietz et al. 1989).

De Guise et al. (1994) and Martineau et al. (1994) found pathological disorders in beluga whales (*D. leucas*) in the St. Lawrence River and they associated these effects with exposure of the whales to PAHs and PCBs.

6.2 Endocrine Disruption in Freshwater Mammals

In some studies, it was observed that different otter species were affected by EDCs.

A study on river otter (*L. Canadensis*) from the Columbia River in the US showed that abnormalities such as reduced baculum length and weight and aspermatogenesis resulted from delayed development. The delayed development was thought to result from exposure to organochlorine insecticides, PCBs, dioxins and furans (Henny et al. 1996).

The Department of Environmental Conservation in New York reported an increase of the river otter population in four New York State counties between 1960 and 1970. The increase within the river otter population at that time was attributed to a reduced exposure to organochlorines from improved water quality (Grannis 2008; http://www.dec.state.ny.us/).

The decline of the European otter (*Luta lutra*) population, in Europe, was the subject of many studies. Results of these studies linked the decline to PCB exposure (Brunström et al. 1998; Kihlström et al. 1992; Leonards 1997; Roos et al. 2001). Roos et al. (2001) observed that, in 1990, decreasing PCB concentrations resulted in increases in European otter populations in Sweden. This result confirmed the suspicion that PCBs had been the major cause for the European otter decline during the previous decades.

Declining American mink (*Mustela vison*) populations have occurred in different areas of the Great Lakes USA/Canada. It was shown that fish from the Great Lake region contain high concentrations of numerous synthetic organochlorines, including pesticides and PCBs. The high consumption of contaminated fish by mink was thought to cause their decline in population (Wren 1991). A study by Ankley et al. (1997) confirmed this suspicion. The authors were able to link adverse reproductive outcomes in mink to the high consumption of contaminated fish from the Great Lakes. Giesy et al. (1994a) concluded that PCBs and TCDD (tetrachlorodibenzo-p-dioxin) have the greatest impact on mink populations, compared with other environmental pollutants.

A laboratory study by Brunström et al. (2001) revealed that PCBs affect reproduction success in American mink. Mink exposed to only low concentrations of PCBs over 18 mon, suffered from fetal deaths, abnormalities, decreased survival and decreased growth (Brunström et al. 2001). Reproductive and non-reproductive effects were observed for aquatic mammalian species, and even if the major focus lies on effects on single individuals, there is ample evidence that environmental chemicals can also alter population dynamics.

6.3 Endocrine Disruption in Terrestrial Mammals

The polar bear (*Ursus maritimus*) is a top predator of the Arctic marine ecosystem. Polar bears prey primarily on ringed seals (*P. hispida*) and bearded seals (*Erignathus barbatus*), which live predominantly on sea ice. PCB levels in polar bears are reported to be extremely high. As a result of their diet, the potential for bio-magnification and bio-accumulation of environmental contaminants in this species is high (Bernhoft et al. 1997).

Between 1995 and 1998, male polar bears (*U. maritimus*) from the Svalbard area were investigated for possible endocrine disruption effects caused by organochlorines. Pesticides and PCBs were found to affect bear testosterone concentration, and the continuing presence of these compounds may affect sexual development and reproductive function (Oskam et al. 2003).

Wiig et al. (1998) observed pseudohermaphrodism in female polar bears from Svalbard, Spitsbergen. Some of the observed bears had a 20-mm penis containing a baculum; other bears exhibited aberrant genital morphology and a high degree of clitoral hypertrophy. Pseudohermaphroditism observed in polar bears was thought to be an effect of EDCs. The authors believed that the observed pseudohermaphroditism could be a result of organochlorines, typically PCBs, which concentrate in fat to very high levels.

Cattet (1988) observed incidences of masculinization in female black (*Ursus americanus*) and brown bears (*Ursus arctos*) in Alberta, Canada. Although the cause for this effect is unknown, it was speculated that the examined pseudohermaphrodism was induced by herbicides. Benirschke (1981), on the other hand, suggested that the observed masculinization was caused by endogenous factors (i.e., excessive maternal androgens). It is not now known whether the described effects had any impact at the population level, in these species.

Facemire et al. (1995) were able to show defects of the reproductive, endocrine and immune systems in Florida panthers (*Felix concolor coryi*). The causes for these defects are not known so far. However, it was assumed that environmental chemicals with endocrine disrupting properties may have been responsible.

A study on possible reproductive effects of PCBs on declining European polecat (*Mustela putorius*) populations did not support the hypothesis that PCBs are responsible for the decline of this species in Central Europe. In this report, it was revealed that other environmental factors such as habitat destruction are more likely to have affected this population (Engelhart et al. 2001).

Rodent populations may experience adverse reproductive effects as a result of exposure to environmental chemicals. Linzey and Grant (1994) studied white-footed mice (*Peromyscus leucopus*) inhabiting low-PCB-contaminated woodland. The exposed mice showed a higher population density but greater temporal variability between yr. White-footed mice, (*P. leucopus*) exposed to PCB and cadmium, had significantly lower relative testis weights compared with mice collected from an unpolluted site (Batty et al. 1990). The population of mice exposed to contaminants did not increase, whereas the reference population showed an increase. However, it was not possible to attribute the observed effects to PCBs and/or cadmium, and it is not known if the observed reproductive effects are a result of endocrine disruption. Further research is required to clarify if a link exists in this case or not. A study on Meadow voles (*Microtus pennsylvanicus*) living close to a chemical waste site at the Niagara Falls showed altered population densities, when compared to a reference population. Life expectancy was reduced in exposed voles, and the tissues contained hexachlorocyclohexane and other chlorinated hydrocarbons. These compounds were not found in vole tissues from the reference site and, therefore, it was concluded that these chemicals were responsible for the observed effects (Rowley et al. 1983). Pomeroy and Barrett (1975) observed that the application of carbaryl (a carbamate insecticide) contributed to delayed reproduction and reduced recruitment in cotton rats (*Sigmodon hispidus*).

Organochlorines with endocrine disrupting properties such as PCBs, or other pesticides were found to affect terrestrial mammals in various ways. Some of the observed effects may result in changed population dynamics; however, it was further assumed that environmental chemicals are not the only reason behind population declines in terrestrial mammalian species.

6.4 Impact on Feral Mammals: Conclusion

It is reasonable to believe that mammals have been adversely affected at the population level by environmental contaminants. Research in this area has largely focused on compounds that persist and bio-accumulate in mammals. Despite the strong correlation with organochlorine (e.g., PCBs, PCDFs and PCDDs) exposure and population decline in a number of studies, we still have an incomplete understanding of the specific compounds responsible for observed pathological effects (Reijnders 1999; Troisi and Mason 1998).

Nevertheless, numerous studies have reported symptoms of endocrine disruption or other adverse physiological effects as a result of exposure to substances with endocrine-disrupting properties. In this context, e.g., seals (Reijnders 1986) and mustelids (Kihlström et al. 1992; Leonards 1997; Wren 1991) suffered from increased reproductive effects in the presence of organochlorine chemicals. As a result of EDC exposure, both reproductive and non reproductive dysfunctions were described. However, in the majority of cases, population data are too limited to provide a link between exposure and reproductive outcome, and it is also not known whether induced non-reproductive dysfunctions foster effects at the population level, or not. Furthermore, there is still insufficient knowledge on the impact of other, possibly confounding environmental stressors that act in parallel with pollutants on free-living populations.

7 Summary

A plethora of papers have been published that address the affects of chemicals on wildlife vertebrates. Collectively, they support a connection between environmental pollution and effects on wildlife vertebrate populations; however, causal relationships between exposure, and reproduction or population structure effects have been established for only a few species.

In a vast number of *fish* species, particularly in teleosts, it is accepted that EDCs affect the endocrine system of individuals and may alter sexual development and fertility. However, only few studies have demonstrated population-level consequences as a result of exposure to EDCs. The same applies to fish populations exposed to contaminants or contaminant mixtures with non-endocrine modes of action; few studies link EDCs directly to population affects.

Amphibian populations are declining in many parts of the world. Although environmental chemicals have been shown to affect reproduction and development in single organism tests, the degree to which chemicals contribute to the decline of amphibians, either alone, or in concert with other factors (habitat loss, climate change, introduction of neozoa, UV-B irradiation, and direct exploitation) is still uncertain.

Because *reptilian* endocrinology is so variable among species, EDC effects reported for individual species cannot easily be extrapolated to others. Nevertheless, for some species and locations (e.g., the Lake Apopka alligators), there is considerable evidence that population declines are caused or triggered by chemical pollution.

In *birds*, there is ample evidence for EDC effects on the reproductive system. In some bird species, effects can be linked to population declines (e.g., based on eggshell thinning induced by DDT/DDE). In contrast, other bird species were shown to be rather insensitive to endocrine disruption. Oil spills, which also may exert endocrine effects, are usually regarded to cause only transient bird population effects, although long-term data are largely missing.

Mammal population declines have been correlated with organochlorine pollution. Moreover, numerous studies have attributed reproductive and non-reproductive dysfunctions in mammals to EDC exposure. However, in the majority of cases, it is uncertain if effects at the population level can be attributed to chemical-induced reproductive effects.

Evidence shows that selected species from all vertebrate classes were negatively affected by certain anthropogenic chemicals. Affects on some species are well characterized at the organismal level. However, the proof of a direct link between chemical exposure and population decline was not given for the vast majority of studied species. This review clearly shows the gaps in knowledge that must be filled for the topic area addressed. We, herewith, make a plea for long-term studies designed to monitor effects of various environmental chemicals on wildlife vertebrate populations. Such studies may be augmented or combined with mechanistically-orientated histological, cytological and biochemical parallel investigations, to fill knowledge gaps.

References

Abbott D (2000) (SMV Science Page, Sharks) http://www.ms-starship.com/sciencenew/sharks.htm Posted June 2000. (Viewed Feb 11, 2008).

Adkins EK (1979) Effect of embryonic treatment with estradiol or testosterone on sexual differentiation of the quail brain: Critical period and dose-response relationships. Neuroendocrinology 29:178–185.

Adkins-Regan E, Mansukhani V, Seiwert C, Thompson R (1994) Sexual differentiation of brain and behavior in the Zebra Finch: Critical periods for effects of early estrogen treatment. J Neurobiol 25:865–877.

Agler BA, Kendall SJ, Irons DB, Klosiewski SP (1999) Declines in marine bird populations in Prince William Sound, Alaska coincident with a climatic regime shift. Water Birds 22:98–103.

Allen Y, Scott AP, Matthiessen P, Haworth S, Thain JE, Feist S (1999) Survey of estrogenic activity in United Kingdom estuarine and coastal waters and its effects on gonadal development of the flounder *Platichthys flesus*. Environ Toxicol Chem 18:1791–1800.

Alquezar R, Markich SJ, Booth DJ (2006) Effects of metals on condition and reproductive output of the smooth toadfish in Sydney estuaries, south-eastern Australia. Environ Pollut 142:116–122.

Andersson T, Förlin L, Härdig J, Larsson A (1988) Biochemical and physiological disturbances in fish inhabiting coastal waters polluted with bleached kraft mill effluents. Mar Environ Res 24:233–236.

Ankley GT, Giesy J (1998) Overview of a workshop on screening methods for detecting potential (anti-) estrogenic/androgenic chemicals in wildlife. Environ Toxicol Chem 17:68–87.

Ankley GT, Johnson RD, Toth G, Folmar LC, Detenbeck NE, Bradbury SP (1997) Development of a research strategy for assessing the ecological risk of endocrine disruptors. Rev Toxicol 1:71–106.

Barnthouse LW (1993) Population-level effects. In: Suter GW (ed) Ecological risk assessment. Lewis, Boca Raton, FL, USA, pp. 247–274.

Barrett CJ, Vainio H, Peakall D, Goldstein BD (1997) 12th Meeting of the Scientific Group on Methodologies for the Safety Evaluation of Chemicals: Susceptibility to Environmental Hazards. Environ Health Perspect 105(Suppl 4):699–737.

Barron MG, Galbraith H, Beltman D (1995) Comparative reproductive and developmental toxicology of PCBs in birds. Comp Biochem Physiol C 112:1–14.

Barton HA, Andersen RE (1998) Endocrine active compounds: From biology to dose response assessment. Crit Rev Toxicol 28:363–423.

Batty J, Levitt RA, Biondi N, Polin D (1990) An ecotoxicological study of the white footed mouse (*Peromyscus leucomus*) inhabiting a polychlorinated biphenyls-contaminated area. Arch Environ Contam Toxicol 19:283–290.

Beebee TJR, Griffiths RA (2005) The amphibian decline crisis: A watershed for conservation biology? Biol Conserv 125:271–285.

Benirschke K (1981) Hermaphrodites, freemartins, mosaics, and chirnacras in animals. In: Austin CR, Edwards RG (eds) Mechanisms of Sex Differentiation in Animals and Man, Academic Press, London, pp. 421–463.

Bennett RS, Williams BA, Schmedding DW, Bennett JK (1991) Effects of dietary exposure to methyl parathion on egg laying and incubation in mallards. Environ Toxicol Chem 10:501–507.

Bergman A (1999a) Health condition of the Baltic grey seal (*Halichoerus grypus*) during two decades, gynaecological health improvement but increased prevalence of colonic ulcers. Acta Pathol Microbiol Immunol Scand 107:270–282.

Bergman A (1999b) Prevalence of lesions associated with a disease complex in the Baltic grey seal (*Halichoerus grypus*) during 1977–1996. In: O'Shea TJ, Reeves RR, Long AK (eds)

Marine Mammals and Persistent Ocean Contaminants: Proceedings of the Marine Mammal Commission Workshop, Keystone, Colorado, pp. 139–143.

Bergman A, Olsson M (1985) Pathology of Baltic grey seal and ringed seal females with special reference to adrenocortical hyperplasia: Is environmental pollution the cause of a widely distributed disease syndrome? Finnish Game Res 44: 47–62.

Bergman A, Bergstrand A, Bignert A (2001) Renal lesions in Baltic grey seals (Halichoerus grypus) and ringed seals (*Phoca hispida botnica*). Ambio 30:397–409.

Bernhoft A, Wiig O, Skaare JU (1997) Organochlorines in polar bears (*Ursus maritimus*) at Svalbard. Environ Pollut 95:159–175.

Berrill M, Bertram S, Wilson A, Louis S, Brigham D, Stromberg C (1993) Lethal and sublethal impacts of pyrethroid insecticides on amphibian embryos and tadpoles. Environ Toxicol Chem 12:525–539.

Berrill M, Bertra S, McGillivray L, Kolohon M, Pauli B (1994) Effects of low concentrations of forest-use pesticides on frog embryos and tadpoles. Environ Toxicol Chem 13:657–664.

Berrill M, Coulson D, McGillivray L, Pauli B (1998) Toxicity of endosulfan to aquatic stages of anuran amphibians. Environ Toxicol Chem 17:1738–1744.

Bignert A, Olsson M, de Wit C, Litzen K, Rappe C, Reutergardh L (1994) Biological variation—an important factor to consider in ecotoxicological studies based on environmental samples. Fresenius J Anal Chem 348:76–85.

Bigsby R, Chapin RE, Daston GP, Davis BJ, Gorski J, Gray LE, Howdeshell KL, Zoeller RT, Vom Saal FS (1999) Evaluating the effects of endocrine disrupters on endocrine function during development. Environ Health Perspect 107:613–618.

Bishop CA, Gendron AD (1998) Reptiles and amphibians: The shy and sensitive vertebrates of the Great Lakes basin and St. Lawrence River. Environ Monit Assess 53:225–244.

Bishop CA, Brooks RJ, Carey JH, Ng P, Norstrom RJ, Lean DR (1991) The case for a cause-effect linkage between environmental contamination and development in eggs of the common snapping turtle (*Chelydra s. serpentina*) from Ontario. Can J Toxicol Environ Health 33(4):521–547.

Bishop CA, Ng P, Pettit KE, Kennedy SW, Stegeman JJ, Norstrom RJ, Brooks RJ (1998) Environmental contamination and developmental abnormalities in eggs and hatchlings of the common snapping turtle (*Chelydra serpentina serpentina*) from the Great Lakes St.-Lawrence River basin (1989–91). Environ Pollut 101:143–156.

Boone MD, Semlitsch RD (2001) Interactions of an insecticide with larval density and predation in experimental amphibian communities. Conserv Biol 15:228–238.

Boss WR, Witschi E (1947) The permanent effects of early stilbestrol injections on the sex organs of the herring gull (*Larus argentatus*). J Exp Biol 105:61–77.

Bowler JK (1977) Longevity of reptiles and amphibians in North American collections. Society for the study of amphibians and reptiles, Association of Fish and Wildlife Agencies 37:212–221.

Bridges CM (1997) Tadpole swimming performance and activity affected by acute exposure to sublethal levels of carbaryl. Environ Toxicol Chem 16:1935–1939.

Bridges CM (1999) Predator-prey interactions between two amphibian species: Effects of insecticide exposure. Aquat Ecol 33:205–211.

Bridges CM (2000) Long-term effects of pesticide exposure at various life stages of the southern leopard frog (*Rana sphenocephala*). Arch Environ Contam Toxicol 39:91–96.

Bridges CM, Semlitsch RD (2000) Variation in pesticide tolerance of tadpoles among and within species of ranidae and patterns of amphibian decline. Conserv Biol 14:1490–1499.

Brunström B, Olsson M, Roos A (1998) 2,3,7,8-TCDD equivalent concentrations in livers from Swedish otters determined with a bioassay. Organohalogen Compd 39:149–151.

Brunström B, Lund B-O, Bergman A, Asplund L, Athanassiadis I, Athanassiadou M, Jensen S, Örberg J (2001) Reproductive toxicity in mink (*Mustela vison*) chronically exposed to environmentally relevant polychlorinated biphenyl concentrations. Environ Toxicol Chem 20:2318–2327.

Burkhardt-Holm P, Peter A, Segner H (2002) Decline of fish catch in Switzerland. Project fishnet: A balance between analysis and synthesis. Aquat Sci 64:36–54.

Burn DM, Doroff AM (2005) Decline in sea otter (*Enhydra lutris*) populations along the Alaska Peninsula, 1986–2001. Fishery Bull 103:270–279.

Campana SE, Marks L, Joyce W, Kohler NE (2006) Effects of recreational and commercial fishing on blue sharks (*Prionace glauca*) in Atlantic Canada, with inferences on the North Atlantic population. Can J Fisheries Aquat Sci 63:670–682.

Carls MG, Marty GD, Hose JE (2002) Synthesis of the toxicological impacts of the Exxon Valdez oil spill on Pacific herring (*Clupea pallasi*) in Prince William Sound, Alaska, USA. Can J Fisheries Aquat Sci 59:153–172.

Carson R L (1962) Silent Spring, Houghton Mifflin, Wilmington, MA, USA.

Casillas E, Misitano D, Johnson LL, Rhodes LD, Collier TK, Stein JE, McCain BB, Varanasi U (1991) Inducibility of spawning and reproductive success of female English sole (*Parophrys vetulus*) from urban and nonurban areas of Puget Sound, Washington. Mar Environ Res 31:99–122.

Castelege I, Hemery G, Roux N, D'Elbeee J, Lalanne Y, D'Amico F, Mouches C (2004) Changes in abundance and at-sea distribution of seabirds in the Bay of Biscay prior to, and following the "Erika" oil spill. Aquat Living Res 17:361–367.

Cattet M (1988) Abnormal sex differentiation in black bears (*Ursus americanus*) and brown bears (*Ursus arctos*). J Mammal 69:849–852.

Cobb GP, Wood PD (1997) PCB concentrations in eggs and chorioallantoic membranes of loggerhead sea turtles (*Caretta caretta*) from the Cape Romain National Wildlife Refuge. Chemosphere 34:539–549.

Congdon JD, Tinkle DW, Breitenbach GW, van Loben Sels RC (1983) Nesting ecology and hatching success in the turtle *Emydoidea blandingii*. Herpetologica 39:417–429.

Conover MR, Hunt GL Jr. (1984a) Experimental evidence that female-female pairs in gulls result from a shortage of breeding males. Condor 86:472–476.

Conover MR, Hunt GL Jr. (1984b) Female-female pairing and sex ratios in gulls: An historical perspective. Wilson Bull 96:619–625.

Conover MR, Miller DE, Hunt GL Jr. (1979) Female-female pairs and other unusual reproductive associations in ring-billed and California gulls. The Auk 96:6–9.

Couture P, Rajotte JW (2003) Morphometric and metabolic indicators of metal stress in wild yellow perch (Perca flavescens) from Sudbury, Ontario, A review. J Environ Monit 5:216–221.

Crain DA, Guillette LJ Jr. (1998) Reptiles as models of contaminant-induced endocrine disruption. Anim Reprod Sci 53:77–86.

Crews D, Bergeron JM, McLachlan JA (1995) The role of estrogen in turtle sex determination and the effect of PCBs. Environ Health Perspect 103(Suppl 7):73–77.

Davidson C (2004) Declining downwind: Amphibian population declines in California and historical pesticide use. Ecol Appl 14:1892–1902.

Davidson C, Shaffer HB, Jennings MR (2001) Declines of the California red-legged frog: Climate, UV-B, habitat, and pesticides hypotheses. Ecol Appl 11:464–479.

De Guise S, Lagace A, Beland P (1994) True hermaphroditism in a St. Lawrence beluga whale (*Delphinapterus leucas*). J Wildl Dis 30:287–290.

De Lafontaine Y, Gilbert NL, Dumouchel F, Brochu C, Moore S, Pelletier E, Dumont P, Branchaud A (2002) Is chemical contamination responsible for the decline of the copper redhorse (*Moxostoma hubbsi*), an endangered fish species, in Canada? Sci Tot Environ 298:25–44.

De Swart RL, Ross PS, Vedder LJ, Timmerman HH, Heisterkamp SH, Van Loveren H, Vos JG, Reijnders PJH, Osterhaus ADME (1994) Impairment of immune functions in harbour seals (*Phoca vitulina*), feeding on fish from polluted costal waters. Ambio 23:155–159.

De Swart RL, Harder TC, Ross PS, Vos HW, Osterhaus ADME (1995) Morbilliviruses and morbillivirus diseases of marine mammals. Infect Agents Dis 4:125–130.

Dean TA, Bodkin JL, Jewett SC, Monson DH, Jung D (2000) Changes in sea urchins and kelp following a reduction in sea otter density as a result of the Exxon Valdez oil spill. Mar Ecol Prog Ser 199:281–291.

DeLong RL, Gilmartin WG, Simpson JG (1973) Premature births in Californian sealions: Association with high organochlorine pollutant residue levels. Science 181:1168–1170.
Dietz R, Heide-Jorgensen MP, Härkönen T (1989) Mass deaths of harbour seals (*Phoca vitulina*) in Europe. Ambio 18:258–264.
Dwivedi J, Trombetta LD (2006) Acute toxicity and bioaccumulation of tributyltin in tissues of *Urolophus jamaicensis* (yellow stingray). J Toxicol Environ Health A 69:1311–1323.
Elliott JE, Norstrom RJ, Keith JA (1988) Organochlorines and eggshell thinning in northern gannets (*Sula bassanus*) from eastern Canada. Environ Pollut 52:81–102.
Engelhart A, Behnisch P, Hagenmaier H, Apfelbach R (2001) PCBs and their putative effects on polecat (*Mustela putorius*) populations in Central Europe. Ecotoxicol Environ Saf 48:178–182.
Esler D, Bowman TD, Trust KA, Ballachey BE, Dean TA, Jewett SC, O'Clair CE (2002) Harlequin duck population recovery following the 'Exxon Valdez' oil spill: Progress, process and constraints. Mar Ecol Prog Ser 241:271–286.
Evans DO, Warren GJ, Cairns VW (1990) Assessment and management of fish community health in the Great Lakes: Synthesis and recommendations. J Great Lakes Res 16:639–669.
Facemire CF, Gross TS, Guillette LJJ (1995) Reproductive impairment in the Florida panther: Nature or nurture? Environ Health Perspect 103:79–86.
Fair PA, Becker PR (2000) Review of stress in marine mammals. J Aquat Ecosyst Stress Recov 7:335–354.
Fent K (2003) Ökotoxikologie, 2. Auflage. Georg Thieme, Stuttgart, Germany.
Fentress JA, Steele SL, Bart HL Jr., Cheek AO (2006) Reproductive disruption in wild longear sunfish (*Lepomis megalotis*) exposed to kraft mill effluent. Environ Health Perspect 114:40–45.
Fitch MA, Shugart GW (1983) Comparative biology and behaviour of monogamous pairs and one male-two female trios of herring gulls. Behav Ecol Sociobiol 14:1–7.
Folmar LC, Denslow ND, Rao V, Chow M, Crain DA, Enblom J, Marcino J, Guillette LJ (1996) Vitellogenin induction and reduced serum testosterone concentrations in feral male carp (*Cyprinus carpio*) captured near a major metropolitan sewage treatment plant. Environ Health Perspect 104:1096–1101.
Fox G, Gilman A, Peakall D, Anderka F (1978) Behavioral abnormalities of nesting Lake Ontario herring gulls. J Wild Manag 42:477–483.
Fox GA (1991) Practical causal inference for ecoepidemiologists. J Toxicol Environ Health 33:359–373.
Fox GA (1992) Epidemiological and pathobiological evidence of contaminant-induced alterations in sexual development in free-living wildlife. In: Colborn T, Clement C (eds) Chemically-Induced Alterations in Sexual and Functional Development: The Wildlife/Human Connection. Princeton Scientific Publishing Co., Princeton, NJ, USA, pp. 147–158.
Fox GA (1993) What have biomarkers told us about the effects of contaminants on the health of fish-eating birds in the Great Lakes? The theory and literature review. J Great Lakes Res 19:722–736.
Fox GA, MacCluskie MC, Brook RW (2005) Are current contaminant concentrations in eggs and breeding female Lesser Scaup of concern? Condor 107:50–61.
Freeman JL, Beccue N, Rayburn AL (2005) Differential metamorphosis alters the endocrine response in anuran larvae exposed to T3 and atrazine. Aquat Toxicol 75:263–276.
Frost KJ, Lowry LF, Ver Hoef JM (1999) Monitoring the trend of harbor seals in Prince William Sound, Alaska, after the Exxon Valdez oil spill. Marine Mammal Sci 15:494–506.
Fry DM (1995) Reproductive effects in birds exposed to pesticides and industrial chemicals. Environ Health Perspect 103:165–171.
Fry DM, Toone CK (1981) DDT-induced feminization of gull embryos. Science 213:922–924.
Fry DM, Toone CK, Speich SM, Peard RJ (1987) Sex ratio skew and breeding patterns of gulls: Demographic and toxicological considerations. Stud Avian Biol 10:26–43.
Gagnon MM, Bussieres D, Dodson JJ, Hodson PV (1995) White sucker (*Catostomus commersoni*) growth and sexual maturation in pulp mill-contaminated and reference rivers. Environ Toxicol Chem 14:317–327.

Garshelis DL, Johnson CB (2001) Sea otter population dynamics and the Exxon Valdez oil spill: Disentangling the confounding effects. J Appl Ecol 38:19–35.

Gelsleichter J, Manire CA, Szabo NJ, Cortes E, Carlson J, Lombardi-Carlson L (2005) Organochlorine concentrations in bonnethead sharks (*Sphyrna tiburo*) from four Florida estuaries. Arch Environ Contam Toxicol 48:474–483.

Gibbons JW, Semlitsch RD (1982) Survivorship and longevity of a long-lived vertebrate species: How long do turtles live? J Anim Ecol 51:523–527.

Giesy JP, Ludwig JP, Tillitt DE (1994a) Dioxins, dibenzofurans, PCBs and colonial, fish-eating water birds. In: Schecter A (ed) Dioxin and Health. Plenum Press, New York, USA, pp. 254–307.

Giesy JP, Ludwig JP, Tillitt DE (1994b) Deformities in birds of the Great Lakes region. Environ Sci Technol 28:128–135.

Gilbertson M (1983) Etiology of chick edema disease in herring gulls in the lower Great Lakes. Chemosphere 12:357–370

Gilbertson M, Fox G, (1977) Pollutant-associated embryonic mortality of Great Lakes herring gulls. Environ Pollut 12:211–216.

Gilbertson M, Kubiak TJ, Ludwig JP, Fox G (1991) Great Lakes embryo mortality, edema, and deformities syndrome (GLEMEDS) in colonial fish-eating birds: Similarity to chick edema disease. J Toxicol Environ Health 33:455–520.

Gill AB, Kimber JA (2005) The potential for cooperative management of elasmobranchs and offshore renewable energy development in UK waters. J Mar Biol Assoc UK 85: 1075–1081.

Gill ME, Spiropoulos J, Moss C (2002) Testicular structure and sperm production in flounders from a polluted estuary: A preliminary study. J Exp Mar Biol Ecol 281:41–51.

Gilman A, Peakall D, Hallett D, Fox G, Norstrom R (1979) Herring gulls (*Larus argentatus*) as monitors of contamination in the Great Lakes. In: Animals as Monitors of Environmental Pollution. Symposium on Pathobiology of Environmental Pollutants. National Academy of Sciences, Washington, D.C., pp. 280–289.

Gilmartin WG, DeLong RL, Smith AW, Sweeney JC, De Lappe BW, Risebrough RW, Griner LA, Dailey MD, Peakall DB (1976) Premature parturition in the Californian sea lion. J Wildl Dis 12:104–115.

Grannis P (2008) (New York State Department of Environmental Conservation, P. Grannis, Commissioner) http://www.dec.state.ny.us/ (Viewed Feb 11, 2008).

Grasman K, Scanlon P, Fox G (1998) Reproductive and physiological effects of environmental contaminants in fish-eating birds of the Great Lakes: A review of historical trends. Environ Monit Assess 53:117–145.

Guillette LJ Jr., Edwards TM (2005) Is nitrate an ecologically relevant endocrine disruptor in vertebrates? Integrat Comp Biol 45:19–27.

Guillette LJ Jr., Gross TS, Masson GR, Matter JM, Percival HF, Woodward AR (1994) Developmental abnormalities of the gonad and abnormal sex hormone concentrations in juvenile alligators from contaminated and control lakes in Florida. Environ Health Perspect 102:680–688.

Guillette LJ Jr., Brock JW, Rooney AA, Woodward AR (1999) Serum concentrations of various environmental contaminants and their relationship to sex steroid concentrations and phallus size in juvenile American alligators. Arch Environ Contam Toxicol 36:447–455.

Gutleb AC, Appelman J, Bronkhorst MC, Van Den Berg JHJ, Spenkelink A, Brouwer A, Murk AJ (1999) Delayed effects of pre- and early-life time exposure to polychlorinated biphenyls on tadpoles of two amphibian species (*Xenopus laevis* and *Rana temporaria*). Environ Toxicol Pharmacol 8:1–14.

Hall RJ, Henry PFP (1992) Assessing effects of pesticides on amphibians and reptiles: status and needs. Herpetol J 2:65–71.

Hansen JA, Lipton J, Welsh PG (2002) Relative sensitivity of bull trout (*Salvelinus confluentus*) and rainbow trout (*Oncorhynchus mykiss*) to acute copper toxicity. Environ Toxicol Chem 21:633–639.

Hansen P-D, von Westernhagen H, Rosenthal H (1985) Chlorinated hydrocarbons and hatching success in Baltic herring spring spawners. Mar Environ Res 15:59–76.

Harries JE, Sheahan DA, Jobling S, Matthiessen P, Neall P, Routledge E, Rycroft R, Sumpter JP, Tylor T (1996) A survey of estrogenic activity in United Kingdom inland waters. Environ Toxicol Chem 15:1993–2002.

Harries JE, Sheahan DA, Jobling S, Matthiessen P, Neall P, Sumpter JP, Tylor T, Zaman N (1997) Estrogenic activity in five United Kingdom rivers detected by measurement of vitellogenesis in caged male trout. Environ Toxicol Chem 16:534–542.

Hashimoto S, Bessho H, Hara A, Nakamura M, Iguchi T, Fujita K (2000) Elevated serum vitellogenin levels and gonadal abnormalities in wild male flounder (*Pleuronectes yokohamae*) from Tokyo Bay, Japan. Mar Environ Res 49:37–53.

Hayes TB, Collins A, Lee M, Mendoza M, Noriega N, Stuart AA, Vonk A (2002) Hermaphroditic, demasculinized frogs after exposure to the herbicide atrazine at low ecologically relevant doses. Proc Natl Acad Sci USA 99:5476–5480.

Hecnar S J (1995) Acute and chronic toxicity of ammonium nitrate fertilizer to amphibians from southern Ontario. Environ Toxicol Chem 14:2131–2137.

Henny CJ, Grove RA, Hedstrom OR (1996) A field evaluation of mink and otter on the lower Columbia River and the influence of environmental contaminants, Final Report to the Lower Columbia River Bi-State Water Quality Program (Portland, OR). National Biological Service, Forest and Rangeland Ecosystem Science Center, Corvalis, OR.

Henny CJ, Beal KF, Bury RB, Goggans R (2003) Organochlorine pesticides, PCBs, trace elements and metals in western pond turtle eggs from Oregon. Northwest Sci 77:46–53.

Hilborn R, Eggers D (2000) A review of the hatchery programs for pink salmon in Prince William Sound and Kodiak Island, Alaska. Transact Am Fisheries Soc 129:333–350.

Hodson PV, McWhirter M, Ralph K, Gray B, Thiverge D, Carey J, Van Der Kraak GJ, McWhittle D, Levesque M, (1992) Effects of bleached kraft mill effluent on fish in the St. Maurice River, Quebec. Environ Toxicol Chem 11:1635–1651.

Houlihan JE, Fidlay CS, Schmidt BR, Meyer AH, Kuzmin SL (2000) Quantitative evidence for global amphibian population declines. Nature (Lond) 404:752–755.

Hunt GL Jr., Wingfield JC, Newman A, Farner DS (1980) Sex ratio of Western Gulls on Santa Barbara Island, California. The Auk 97:473–479.

ICES (1992) Report of the Study Group on Seals and Small Cetaceans in Northern European Seas, International Council for the Exploration of the Sea (ICES) CM 1993/N:3.

Irons DB, Kendall SJ, Erickson WP, McDonald LL, Lance BK (2000) Nine years after the Exxon Valdez oil spill: Effects on marine bird populations in Prince William Sound, Alaska. Condor 102:723–737.

Jobling S, Tyler CR (2003) Endocrine disruption in wild freshwater fish. Pure Appl Chem 75:2219–2234.

Jobling S, Sheahan DA, Osborne JA, Matthiessen P, Sumpter JP (1996) Inhibition of testicular growth in rainbow trout (*Oncorhynchus mykiss*) exposed to estrogenic alkylphenolic chemicals. Environ Toxicol Chem 15:194–202.

Jobling S, Nolan M, Tyler CR, Brighty G, Sumpter JP (1998a) Widespread sexual disruption in wild fish. Environ Sci Technol 32:2498–2506.

Jobling S, Tyler CR, Nolan M, Sumpter JP (1998b) The identification of oestrogenic effects in wild fish. London, UK Environment Agency, R and D Technical Report W119.

Johnson LL, Stein JE, Collier TK, Casillas E, McCain B, Varanasi U (1992) Bioindicators of contaminant exposure, liver pathology, and reproductive development in prespawning female winter flounder (*Pleuronectes americanus*) from urban and nonurban estuaries on the northeast Atlantic coast. Washington, D.C., U.S., Department of Commerce, National Oceanic and Atmospheric Administration, NOAA Technical Memorandum NMFSNWFSC-1, 76 pp.

Johnson LL, Sol SY, Ylitalo GM, Hom T, French B, Olson OP, Collier TK (1997) Precocious sexual maturation and other reproductive anomalies in English sole from an urban waterway, International Council for the Exploration of the Sea, Copenhagen, ICES CM 1997/U:07, 15 pp.

Johnson LL, Misitano D, Sol SY, Nelson GM, French B, Ylitalo GM, Hom T (1998) Contaminant effects on ovarian development and spawning success in rock sole from Puget Sound, Washington. Transact Am Fisheries Soc 127:375–392.

Johnstone R, Court G, Fesser A, Bradley D, Oliphant L, MacNeil J (1996) Long-term trends and sources of organochlorine contamination in Canadian tundra peregrine falcons, *Falco peregrinus tundrius*. Environ Pollut 93:109–120.

Kihlström JE, Olsson M, Jensen S, Johansson A, Ahlbom J, Bergman A (1992) Effects of PCB and different fractions of PCB on the reproduction of the mink (*Mustela vison*). Ambio 21:563–569.

Kime DE (1998) Endocrine Disruption in Fish. Kluwer Academic Publishers, Boston, USA.

Kime DE, Nash JP, Scott AP (1999) Vitellogenesis as a biomarker of reproductive disruption by xenobiotics. Aquaculture 177:345–352.

Kloas W (2002) Amphibians as a model for the study of endocrine disruptors. Int Rev Cyt 216:1–57.

Kostamo A, Hyvärinen H, Pellinen J, Kukkonen JVK (2002) Organochlorine concentrations in the Saimaa ringed seal (*Phoca hispida saimensis*) from Lake Haukivesi, Finland, 1981 to 2000, and in its diet today. Environ Toxicol Chem 21:1368–1376.

Kovacs TG, Gibbons JS, Tremblay LA, O'Connor BI, Martel PH, Voss RI (1995) The effect of a secondary-treated bleached kraft pulpmill effluent on aquatic organisms as assessed by short-term and long-term laboratory tests. Ecotoxicol Environ Saf 31:7–22.

Lamb T, Bickham JW, Lyne TB, Gibbons JW (1995) The slider turtle as an environmental sentinel: Multiple tissue assays using flow cytometric analysis. Ecotoxicology 4(1):5–13.

Lance VA (1994) Introduction: Environmental sex determination in reptiles: Patterns and processes. J Exp Zool 270:1–2.

Leonards PEG (1997) PCBs in mustelids: Analysis, food chain transfer and critical levels. Ph.D. Thesis, Free University of Amsterdam, The Netherlands.

Linzey AV, Grant DM (1994) Characteristics of a white footed mouse (*Peromyscus leucopus*) population inhabiting polychlorinated biphenyls contaminated site. Arch Environ Contamin Toxicol 27:521–526.

Ludwig JP (1984) Decline, resurgence and population dynamics of Michigan and Great Lakes double-crested cormorants. Jack-Pine Warbler 62:91–102.

Ludwig JP, Giesy JP, Summer CL, Bowerman WW, Aulerich R, Bursian S, Auman HJ, Jones PD, Williams LL, Tillitt DE, Gilbertson M (1993) A comparison of water quality criteria for the Great Lakes based on human and wildlife health. J Great Lakes Res 19:789–807.

Lund B-O (1994) *In vitro* adrenal bioactivation and effects on steroid metabolism of DDT, PCBs and their metabolites in the grey seal (*Halichoerus grypus*). Environ Toxicol Chem 13:911–917.

Lye CM, Frid CJJ, Gill ME, McCormick D (1997) Abnormalities in the reproductive health of flounder *Platichthys flesus* exposed to effluent from a sewage treatment works. Mar Pollut Bull 34:34–41.

Lye CM, Frid CLJ, Gill ME (1998) Seasonal reproductive health of flounder *Platichthys flesus* exposed to sewage effluent. Mar Ecol Prog Ser 170:249–260.

Lye CM, Frid CLJ, Gill ME, Cooper DW, Jones DM (1999) Estrogenic alkylphenols in fish tissues, sediments and waters from the U.K. Tyne and Tees estuaries. Environ Sci Technol 33:1009–1014.

Mac MJ, Schwartz TR, Edsall CC, Frank AM (1993) Polychlorinated biphenyls in Great Lakes trout and their eggs: Relations to survival and congener composition 1979–1988. J Great Lakes Res 19:752–765.

Maki AW, Brannon EJ, Gilbertson LG, Moulton LL, Skalski JR (1995) An assessment of oil-spill effects on pink salmon populations following the Exxon Valdez oil spill—Part 2: Adults and escapement. In: Wells PG, Butler JN, Hughes JS (eds) *Exxon Valdez Oil Spill: Fate and Effects in Alaskan Waters ASTM Special Technical Publication # 1219*, American Society for Testing and Materials, Phildaelphia, PA, USA, pp. 585–625.

Maltby L (1999) Studying stress: The importance of organism-level responses. Ecol Appl 9:431–440.

Marco A, Quilchano C, Blaustein AR (1999) Sensitivity to nitrate and nitrite in pond-breeding amphibians from the Pacific Northwest, USA. Environ Toxicol Chem 18:2836–2839.

Martineau D, De Guise S, Fournier M, Shugart L, Girard C, Lagace A, Beland P (1994) Pathology and toxicology of beluga whales from the St. Lawrence Estuary, Québec, Canada, Past, present and future. Sci Tot Environ 154:201–215.

Matthiessen P, Gibbs PE (1998) Effects on fish of estrogenic substances in English rivers. In: Kendall R, Dickerson J, Giesy J, Suk W (eds) Principles and Processes for Evaluating Endocrine Disruption in Wildlife, SETAC Press, Pensacola, FL, USA, pp. 239–247.

McArthur MLB, Fox GA, Peakall DB, Philogene BJR (1983) Ecological significance of behavioral and hormonal abnormalities in breeding Ring Doves fed an organochlorine chemical mixture. Arch Environ Contamin Toxicol 12:343–353.

Mellanen P, Soimasuo M, Holmbom B, Oikari A, Santti R (1999) Expression of the vitellogenin gene in the liver of juvenile whitefish (*Coregonus lavaretus* L. *s.l.*) exposed to effluents from pulp and paper mills. Ecotox Environ Saf 43:133–137.

Mikkelsen M, Jenssen BM (2006) Polychlorinated biphenyls, sex steroid hormones and liver retinoids in adult male European common frogs *Rana temporaria*. Chemosphere 63:707–715.

Milnes MR, Allen D, Bryan TA, Sedacca CD, Guillette LJ Jr. (2004) Developmental effects of embryonic exposure to toxaphene in the American alligator (*Alligator mississippiensis*). Comp Biochem Physiol C 138:81–87.

Minier C, Levy F, Rabel D, Bocquene G, Godefroy D, Burgeot T, Leboulenger F (2000) Flounder health status in the Seine Bay, a multibiomarker study. Mar Environ Res 50:373–377.

Monod G (1985) Egg mortality of Lake Geneva charr (*Salvelinus alpinus* L.) contaminated with PCB and DDT derivatives. Bull Environ Contamin Toxicol 35:531–536.

Munkittrick KR, Portt CB, Van Der Kraak GJ, Smith IR, Rokosh DA (1991) Impact of bleached kraft mill effluent on population characteristics, liver MFO activity, and serum steroid levels of a Lake Superior white sucker (*Catastomus commersoni*) population. Can J Fisheries Aquat Sci 48:1371–1380.

Munkittrick KR, Van Der Kraak GJ, McMaster ME, Portt CB (1992a) Response of hepatic MFO activity and plasma sex steroids to secondary treatment of bleached kraft pulp mill effluent and mill shutdown. Environ Toxicol Chem 11:1427–1439.

Munkittrick KR, Van Der Kraak G, McMaster M, Portt C (1992b) Reproductive dysfunction and MFO activity in three species of fish exposed to bleached kraft mill effluent at Jackfish Bay, Lake Superior. Wat Pol Res J Can 27:439–446.

Munkittrick KR, McMaster, ME, McCarthy L, Servos M, Van Der Kraak G (1998) An overview of recent studies on the potential of pulp-mill effluents to alter reproductive parameters in fish. J Toxicol Environ Health B 1:347–371.

Nichols J, Bradbury S, Swartout J (1999) Derivation of wildlife values for mercury. J Toxicol Environ Health B 2:325–355.

Nisbet ICT, Fiy DM, Hateh JJ, Lynn B (1996) Feminization of male common tern embryos is not correlated with exposure to specific PCB congeners. Bull Environ Contamin Toxicol 57:895–901.

NRC (1999) Hormonally Active Agents in the Environment. Washington, D.C., National Research Council (USA), National Academy Press, 414 pp.

Oaks JL, Gilbert M, Virani MZ, Watson RT, Meteyer CU, Rideout BA, Shivaprasad HL, Ahmed S, Chaudhry MJI, Arshad M, Mahmood S, Ali A, Khan AA (2004) Diclofenac residues as the cause of vulture population decline in Pakistan. Nature (Lond) 427:630–633.

Olafsson PG, Bryan AM, Bush B, Stone W (1983) Snapping turtles: A biological screen for PCBs. Chemosphere 12:1525–1523.

Oskam IC, Ropstad E, Dahl E, Lie E, Derocher AE, Wiig Ø, Larsen S, Wiger R, Skaare JU (2003) Organochlorines affect the major androgenic hormone, testosterone, in male polar bears (*Ursus maritimus*) at Svalbard. J Toxicol Environ Health A 66:2119–2139.

Palmer BD, Guillette LJ Jr. (1988) Histology and functional morphology of the female reproductive tract of the tortoise *Gopherus polyphemus*. Am J Anat 183:200–211.

Palmer BD, Guillette LJ Jr. (1990) Morphological changes in the oviductal endometrium during the reproductive cycle of the tortoise, *Gopherus polyphemus*. J Morphol 204:323–333.

Palmer BD, Guillette LJ Jr. (1992) Alligators provide evidence for the evolution of an archosaurian mode of oviparity. Biol Reprod 46:39–47.

Palmer BD, Perkins MJ, Massie K, Simon MS, Uribe UCA (1997) Reproductive anatomy and physiology of reptiles: Evolutionary and ecological perspectives. In: Ackerman L (ed) The Biology, Husbandry and Health Care of Reptiles and Amphibians, Vol. III. T.F.H. Publications, Neptune City, NJ, USA, pp. 54–87.

Peakall DB (1986) Accumulation and effects on birds. In: Waid JS (ed) PCBs in the Environment, Vol. II, CRC Press, Boca Raton, FL, USA, pp. 31–47.

Peakall DB (1988) Known effects of pollutants on fish-eating birds in the Great Lakes of North America. In: Schmidtke NW (ed) Toxic Contamination in Large Lakes, Vol. II. Chronic Effects of Toxic Contaminants in Large Lakes, Lewis Publishers, Chelsea, MI, USA, pp. 39–54.

Peakall DB, Peakall ML (1973) Effect of a polychlorinated biphenyl on the reproduction of artificially and naturally incubated dove eggs. J Appl Ecol 10:863–868.

Pechmann JHK, Scott DE, Semlitsch RD, Caldwell JP, Vitt LJ, Gibbons JW (1991) Declining amphibian populations: The problem of separating human impacts from natural fluctuations. Science 253:892–895.

Pechmann, JHK, Wilbur HM (1994) Putting declining amphibian populations in perspective: Natural fluctuations and human impacts. Herpetologica 50:65–84.

Petersen GI, Gerup J, Nilsson L, Larsen JR, Schneider R (1997) Body burdens of lipophilic xenobiotics and reproductive success in Baltic cod (*Gadus morhua* L.). International Council for the Exploration of the Sea, Copenhagen, ICES CM 1997/U, 10, 22 pp.

Pierotti R (1981) Male and female parental roles in the western gull under different environmental conditions. The Auk 98:532–549.

Pomeroy SE, Barrett GW (1975) Dynamics of enclosed small mammal populations in relation to an experimental pesticide application. Am Midl Nat 93:91–106.

Portelli MJ, De Solla SR, Brooks RJ, Bishop CA (1999) Effect of dichlorodiphenyltrichloroethane on sex determination of the common snapping turtle (*Chelydra serpentina serpentina*). Ecotox Environ Saf 43:284–291.

Pough FH, Janis CM, Heiser JB (2004) Vertebrate Life, 7th ed. Prentice Hall, Upper Saddle River, NJ, USA.

Price IM, Weseloh DV (1986) Increased numbers and productivity of double-crested cormorants (*Phalacrocorax auritus*) on Lake Ontario. Can Field Nat 100:474–482.

Purdom CE, Hardiman PA, Bye VJ, Eno NC, Tyler CR, Sumpter JP (1994) Estrogenic effects of effluents from sewage treatment works. Chem Ecol 8:275–285.

Reeder AL, Ruiz MO, Pessier A, Brown LE, Levengood JM, Phillips CA, Wheeler MB, Warner R, Beasley VR (2005) Intersexuality and the cricket frog decline: Historic and geographic trends. Environ Health Perspect 113:261–265.

Reijnders PJH (1980) Organochlorine and heavy metal residues in harbour seals from the Wadden Sea and their possible effects on reproduction. Neth J Sea Response 14:30–65.

Reijnders PJH (1986) Reproductive failure in common seals feeding on fish from polluted coastal waters. Nature (Lond) 324:456–457.

Reijnders PJH (1990) Progesterone and oestradiol-17 -concentration profiles throughout the reproductive cycle in harbour seals (*Phoca vitulina*). J Reprod Fertil 90:403–409.

Reijnders PJH (1999) Reproductive and developmental effects of endocrine disrupting chemicals on marine mammals. In: O'Shea TJ, Reeves RR, Long KA (eds) Marine mammals and Persistent Ocean Contaminants: Proceedings of the Marine Mammal Commission Workshop. Keystone, Colorado, 12–15 October 1998, pp. 139–143.

Relyea RA (2004) Synergistic impacts of malathion and predatory stress on six species of North American tadpoles. Environ Toxicol Chem 23:1080–1084.

Relyea RA, Mills N (2001) Predator-induced stress makes the pesticide carbaryl more deadly to gray treefrog tadpoles (*Hyla versicolor*). Proc Natl Acad Sci USA 98:2491–2496.

Rijnsdorp AD, Vethaak AD (1997) Changes in reproductive parameters of North Sea plaice and sole between 1960 and 1995. International Council for the Exploration of the Sea, Copenhagen, ICES C.M. 1997/U, 14, 9 pp.

Robinson RD, Carey JH, Solomon KR, Smith IR, Servos MR, Munkittrick KR (1994) Survey of receiving-water environmental impacts associated with discharges from pulp mills. 1. Mill characteristics, receiving-water chemical profiles and lab toxicity tests. Environ Toxicol Chem 13:1075–1088.

Rogan WJ, Gladen BC, Hung KL, Koong SL, Shih LY, Taylor JS, Wu YC, Yang D, Ragan NB, Hsu CC (1988) Congenital poisoning by polychlorinated biphenyls and their contaminants in Taiwan. Science 241:334–336.

Rohr JR, Palmer BD (2005) Aquatic herbicide exposure increases salamander desiccation risk eight months later in a terrestrial environment. Environ Toxicol Chem 24:1253–1258.

Roos A, Bergman A, Greyerz E, Olsson M (1998) Time trend studies on DDT and PCB in juvenile grey seals (*Halichoerus grypus*), fish and guillemot eggs from the Baltic Sea. Organochlorine Compd 39:109–112.

Roos A, Greyerz E, Olsson M, Sandegren F (2001) The otter (*Lutra lutra*) in Sweden—Population trends in relation to DDT and total PCB concentrations during 1968–1998. Environ Pollut 111:457–469.

Rowley MH, Christian JJ, Basu DK, Pawlikowski MA, Paigen B (1983) Use of small mammals (voles) to assess a hazardous waste site at Love Canal, Niagara Falls, New York. Arch Environ Contamin Toxicol 12:383–397.

Russell RW, Hecnar SJ, Haffner GD (1995) Organochlorine pesticide residues in southern Ontario spring peepers. Environ Toxicol Chem 14:815–817.

Russell RW, Gillan KA, Haffner GD (1997) Polychlorinated biphenyls and chlorinated pesticides in southern Ontario, Canada, green frogs. Environ Toxicol Chem 16:2258–2263.

Sandström O, Neuman E, Karas P (1988) Effects of a bleached pulp mill effluent on growth and gonad function in Baltic coastal fish. Water Sci Technol 20:107–118.

Schmidt CW (1997) Amphibian deformities continue to puzzle researchers. Environ Sci Technol 31:324A–326A.

Shugart GW (1980) Frequency and distribution of polygamy in Great Lakes herring gulls in 1978. Condor 82:426–429.

Storelli MM, Storelli A, Marcotrigiano GO (2005) Concentrations and hazard assessment of polychlorinated biphenyls and organochlorine pesticides in shark liver from the Mediterranean Sea. Mar Pollut Bull 50:850–855.

Storrs SI, Kiesecker JM (2004) Survivorship patterns of larval amphibians exposed to low concentrations of atrazine. Environ Health Perspect 112:1054–1057.

Stott GG, Haensly WE, Neff JM, Sharp JR (1983) Histopathologic survey of ovaries of plaice, *Pleuronectes platessa* L., from Aber Wrach and Aber Benoit, Brittany, France: Long-term effects of the *Amoco Cadiz* crude oil spill. J Fish Dis 6:429–437.

Struger J, Weseloh DV (1985) Great Lakes Caspian terns: Egg contaminants and biological implications. Colonial Water Birds 8:142–149.

Struger J, Weseloh DV, Hallett DJ, Mineau P (1985) Organochlorine contaminants in herring gull eggs from the Detroit and Niagara Rivers and Saginaw Bay (1978–1982): Contaminant discriminants. J Great Lakes Res 11:223–230.

Talent LG, Dumont JN, Bantle JA, Janz DM, Talent SG (2002) Evaluation of western fence lizards (*Sceloporus occidentalis*) and eastern fence lizards (*Sceloporus undulatus*) as laboratory reptile models for toxicological investigations. Environ Toxicol Chem 21:899–905.

Tanabe S, Watanabe S, Kan H, Tatsukawa R (1988) Capacity and mode of PCB metabolism in small cetaceans. Mar Mammal Sci 4:103–124.

Teirseire H, Guy V(2000) copper-induced changes in antioxidation enzymes activities in fronds of dockweed (Lemna minor). Plant science 153:65-72.

Theodorakis CW, Rinchard J, Carr JA, Park J-W, Mcdaniel L, Liu F, Wages M (2006) Thyroid endocrine disruption in stonerollers and cricket frogs from perchlorate-contaminated streams in east-central Texas. Ecotoxicology 15:31–50.

Thorne RE (2004) Acoustic surveying of pelagic fish in shallow water. Geoscience and Remote Sensing Symposium (IGARSS) Wiley-IEEE Press, Hoboken, NJ, USA. 2:1426–1429.

Tillitt DE, Ankley GT, Giesy JP, Ludwig JP, Kurita-Matsuba H, Weseloh DV, Ross PS, Bishop CA, Sileo L, Stromborg KL, Larson J, Kubiak TJ (1992) Polychlorinated biphenyl residues and egg mortality in double-crested cormorants from the the Great Lakes. Environ Toxicol Chem 11:1281–1288.

Tori GM, Peterle TJ (1983) Effects of PCBs on mourning dove courtship behaviour. Bull Environ Contamin Toxicol 30:44–49.

Triebskorn R, Köhler H-R (eds) (2001) Double special issue: VALIMAR—VALIdation of biomarkers for the assessment of small stream pollution. J Aquat Ecosyst Stress Recov 8 (Nos. 3–4).

Triebskorn R, Böhmer J, Braunbeck T, Honnen W, Köhler H-R, Lehmann R, Oberemm A, Schwaiger J, Segner H, Schüürmann G, Traunspurger W (2001) The project VALIMAR (Validation of biomarkers for the assessment of small stream pollution): Objectives, experimental design, summary of results, and recommendations for the application of biomarkers in risk assessment. J Aquat Ecosyst Stress Recov 8:161–178.

Triebskorn R, Adam S, Behrens A, Beier S, Böhmer J, Braunbeck T, Casper H, Dietze U, Gernhöfer M, Honnen W, Köhler H-R, Körner W, Konradt J, Lehmann R, Luckenbach T, Oberemm A, Schwaiger J, Segner H, Strmac M, Schüürmann G, Siligato S, Traunspurger W (2003) Establishing causality between pollution and effects at different levels of biological organization: The VALIMAR project. Hum Ecol Risk Assess 9:171–194.

Troisi GM, Mason CF (1998) PCB-associated alteration of hepatic steroid metabolism in harbor seals (*Phoca vitulina*). J Toxicol Environ Health A 61:649–655.

Tyler CR, Routledge EJ (1998) Oestrogenic effects in fish in English rivers with evidence of their causation. Pure Appl Chem 70:1795–1804.

Van Der Kraak G, Hewitt M, Lister A, McMaster ME, Munkittrick KR (2001) Endocrine toxicants and reproductive success in fish. Hum Ecol Risk Assess 7:1017–1025.

Vaughan T, Ryan J, Czaplewski N (2000) Mammalogy, 4th ed., Harcourt College Publishers, Orlando, FL, USA.

Velando A (2005) Population trends and reproductive success of the European shag *Phalacrocorax aristotelis* on the Iberian Peninsula following the Prestige oil spill. J Ornithol 146:116–120.

Von Westernhagen H, Rosenthal H, Dethlefsen V, Ernst W, Harms U, Hansen PD (1981) Bioaccumulating substances and reproductive success in Baltic flounder, *Platichthys flesus*. Aquat Toxicol 1:85–99.

Vos JG, Dybing E, Greim HA, Ladefoged O, Lambre C, Tarazona JV, Brandt I, Vethaak AD (2000) Health effects of endocrine-disrupting chemicals on wildlife, with special reference to the European situation. Crit Rev Toxicol 30:71–133.

Weseloh DVC, Ewins PJ (1994) Characteristics of a rapidly increasing colony of double-crested cormorants (*Phalacrocorax auritus*) in Lake Ontario: Population size, reproductive parameters and band recoveries. J Great Lakes Res 20:443–456.

Weseloh DV, Teeple SM, Gilbertson M (1983) Double-crested cormorants of the Great Lakes: Egg-laying parameters, reproductive failure and contaminant residues in eggs, Lake Huron 1972–1973. Can J Zool 61:427–436.

Wibbels T, Cowan J, Le Boeuf R (1998) Temperature-dependent sex determination in the red-eared slider turtle, *Trachemys scripta*. J Exp Zool 281:409–416.

Wiig *Φ*, Derocher AE, Cronin MM, Skaare JU (1998) Female pseudohermaphrodite polar bears at Svalbard. J Wildl Dis 34:792–796.

Willingham E (2001) Embryonic exposure to low-dose pesticides: Effects on growth rate in the hatchling red-eared slider turtle. J Toxicol Environ Health A 64:257–272.

Willingham EJ (2005) The effects of atrazine and temperature on turtle hatchling size and sex ratios. Front Ecol Environ 3:309–313.

Worldwatch Institute (2003) (Worldwatch Paper #165). http://www.worldwatch.org/node/1763. Posted March 10, 2003. (Viewed Feb 11, 2008).

Wren CD (1991) Cause-effect linkages between chemicals and populations of mink (*Mustela vison*) and otter (*Lutra canadensis*) in the Great Lakes basin. J Toxicol Environ Health 33:549–585.

Wu TH, Canas JE, Rainwater TR, Platt SG, McMurry ST, Anderson TA (2000) Organochlorine contaminants in Morelet's crocodile (*Crocodylus moreletii*) eggs from Belize. Chemosphere 40:671–678.

Biomarkers in Aquatic Plants: Selection and Utility

Richard A. Brain (✉) and Nina Cedergreen

Contents

1 Introduction ... 50
2 Gene Expression ... 52
  2.1 Overview ... 52
  2.2 Quantification ... 69
  2.3 Applications to Evaluate Xenobiotic Stress ... 70
  2.4 Advantages ... 71
  2.5 Disadvantages ... 72
3 Pathway Specific Metabolites ... 72
  3.1 Overview ... 72
  3.2 Quantification ... 73
  3.3 Applications to Evaluate Xenobiotic Stress ... 73
  3.4 Advantages ... 75
  3.5 Disadvantages ... 75
4 Metabolic Enzymes: Phase I Metabolic Enzymes ... 75
  4.1 Overview ... 75
  4.2 Quantification ... 76
  4.3 Applications to Evaluate Xenobiotic Stress ... 77
  4.4 Advantages ... 77
  4.5 Disadvantages ... 78
5 Metabolic Enzymes: Phase II Metabolic Enzymes ... 78
  5.1 Overview ... 78
  5.2 Quantification ... 79
  5.3 Applications to Evaluate Xenobiotic Stress ... 79
  5.4 Advantages ... 80
  5.5 Disadvantages ... 81
6 Phytochelatins ... 81
  6.1 Overview ... 81
  6.2 Quantification ... 82
  6.3 Applications to Evaluate Xenobiotic Stress ... 82
  6.4 Advantages ... 83
  6.5 Disadvantages ... 83

R.A. Brain
Center for Reservoir and Aquatic Systems Research, Department of Environmental Science, Baylor University, Waco, Texas 76798, USA

D.M. Whitacre (ed.) *Reviews of Environmental Contamination and Toxicology,* Vol 198
doi: 10.1007/978-0-387-09646-9,

7 Flavonoids 83
7.1 Overview 83
7.2 Quantification 84
7.3 Applications to Evaluate Xenobiotic Stress 84
7.4 Advantages 85
7.5 Disadvantages 85
8 Stress Proteins (Heat Shock Proteins) 85
8.1 Overview 85
8.2 Quantification 86
8.3 Applications to Evaluate Xenobiotic Stress 87
8.4 Advantages 88
8.5 Disadvantages 88
9 Reactive Oxygen Species and Scavenging Enzymes 89
9.1 Overview 89
9.2 Quantification 89
9.3 Applications to Evaluate Xenobiotic Stress 91
9.4 Advantages 92
9.5 Disadvantages 92
10 Plant Photosynthetic Pigments 92
10.1 Overview 92
10.2 Quantification 93
10.3 Applications to Evaluate Xenobiotic Stress 93
10.4 Advantages 95
10.5 Disadvantages 95
11 Photosynthesis: Chlorophyll Fluorescence and Gas Exchange 95
11.1 Overview 95
11.2 Quantification 96
11.3 Applications to Evaluate Xenobiotic Stress 96
11.4 Advantages 99
11.5 Disadvantages 99
12 Summary 100
References 101

1 Introduction

This is a review and guide for the selection of appropriate biomarkers in aquatic plants to assess specific xenobiotics or groups of xenobiotics in laboratory or field studies. Emphasis is placed on the predictive ability, utility, sensitivity and specificity of each biomarker as biomonitoring agents. There is currently a large degree of ambiguity surrounding the definition, application and terminology of biomarkers in the context of environmental risk assessment. Biomarkers of exposure are defined as any functional measure of exposure that is characterized at a sub-organism level of biological organization (Adams et al. 2001). In contrast, a biomarker of effect (bioindicator) includes biochemical, physiological, or ecological structures or processes that have been correlated or causally linked to biological effects measured at one or more levels of biological organization (McCarty and Munkittrick 1996). The biomarkers covered in this review are considered in the context of either exposure or effect, or both.

It has been suggested that, for environmental toxicology to progress, a battery of biomarkers will be necessary to evaluate chemical hazards (Ernst and Peterson 1994), and that no biomarker can by itself offer a complete solution. Only a multi-parametric approach, including both physiological biomarkers, biomarkers of general stress and more specific biomarkers can adequately contribute to ecotoxicological diagnostics (how precisely an effect can be identified and/or characterized) (Ferrat et al. 2003). However, measuring a suite of biomarkers will only be useful if they are integrated into a mechanistic model with obvious links to fitness (Forbes et al. 2006). Forbes et al. (2006) suggests that biomarkers are potentially useful for testing hypotheses about mechanisms of chemical impacts at different levels of biological organization. Because the targets of protection in ecological risk assessment are usually populations, communities, and ecosystems, but only rarely individuals, the biomarker response must be tightly and consistently linked to responses at these higher strata, particularly if the biomarkers are to be used as effects indicators (Forbes et al. 2006). The ability to use effects at lower biological levels to predict toxicant effects at higher levels requires substantial improvements in understanding how mechanistic processes at each level are functionally integrated, in terms of whole-organism performance (Forbes et al. 2006). Achieving such understanding poses major theoretical challenges.

Compared to animals, the subject of biomarkers in plants is a comparatively less explored with limited examples in the literature (Ernst and Peterson 1994; Ferrat et al. 2003). Notwithstanding, plant biomarkers have demonstrated utility as biomonitoring agents, in addition to their uses in elucidating modes of action. Currently, the effect measures employed for toxicity testing in plants are largely gross acute ones such as growth rate and biomass measurements, dry or wet weight, and visual symptoms, e.g., chlorosis and necrosis (Davy et al. 2001). However, because toxicity in plants is first manifested at the biochemical level, before whole-organism effects are evident, biochemical effect parameters can be early indicators of xenobiotic stress (Davy et al. 2001). Biochemical effects are typically more sensitive, although their environmental relevance and relationship with gross parameters, such as biomass, is not always evident (Davy et al. 2001). Ernst and Peterson (1994) suggest that biomarkers should: go beyond visible and morphological parameters, establish such processes and products of plants that enable early recognition of xenobiotic stress in a dose- or time-dependent manner, and be observable earlier than visible damage. Furthermore, the more specific a xenobiotic stressor affects a metabolic process the more precisely the signal can be perceived (Ernst and Peterson 1994). Thus, ideally, biomarkers should be selected from the events in biochemical or physiological pathways targeted by toxicants (Ernst and Peterson 1994). Robust assessment of the biochemical pathway or target will normally be required to demonstrate biological relevance across the spectrum of biological organization. If the biochemical effect can not be plausibly related to effects at the morphological level it is of questionable value for ecotoxicological diagnostics.

When plants are exposed to a toxicant, they respond at the molecular level, within minutes to hour, by changing gene expression (Akhtar et al. 2005; Jamers et al. 2006). Within this timeframe, metabolic compounds from affected pathways may either accumulate or deplete pools of metabolites. Concomitantly, plants will also rapidly up-regulate their detoxification systems, which can be divided into three phases: oxidation, reduction and/or hydrolysis reactions (phase 1), conjugation (phase 2), and secondary conjugation followed by immobilization, compartmentation and/or metabolization (phase 3) (Kreuz and Martinoia 1999; Kreuz et al. 1996). For metal contaminants, phytochelatins can be formed as a means of sequestration, inactivation/detoxification; this process has also been demonstrated to occur with flavonoids. Plants may synthesize specific stress proteins, which help ameliorate the effects of toxicants on cell protein function. Because plants synthesize organic compounds by capturing and transforming light energy through photosynthesis using a number of finely tuned biochemical pathways, even slight disturbances in the photosynthetic apparatus can rapidly divert excited electrons away from these pathways. Such diversion may give rise to the formation of reactive oxygen species (ROS) and a resultant rise in ROS-scavenging enzymes. ROS formation will also affect the pigment content and composition in the chloroplasts and eventually photosynthetic efficiency, which is the driver for all plant growth. Hence, the effect of toxicant stress on plants can potentially be measured as changes in activity of selected biochemical steps or processes (effect measures). In the following sections, each effect measure is individually presented. In each section, the following are presented: an overview of the biomarker of interest, how it is measured, potential applications (appropriateness to a particular toxicant or group of toxicants), comparative value relative to morphological endpoints and associated advantages and disadvantages. For each biomarker, the quantified response for both the biomarker and a comparative morphological endpoint (if available) are provided in Table 1. Fig. 1 outlines which plant biomarkers are appropriate for evaluating a given environmental stress.

2 Gene Expression

2.1 Overview

Gene expression is a sensitive indicator of toxicant exposure, disease state and cellular metabolism. In principle, the measurement of gene expression levels after chemical exposure can form a "genetic signature" from the pattern of gene expression changes elicited, both *in vitro* and *in vivo*, and provide information about the mechanism of action (Lettieri 2006). Therefore, genomics information may lead to the development of predictive biomarkers of effect that allow for identification of potentially sensitive populations and earlier predictions of adverse effects (USEPA 2004). Almost without exception, gene expression is altered during toxic stress, from either direct or indirect toxicant exposure (Nuwaysir et al. 1999). The challenge is to define, under given experimental conditions, the characteristic and specific pattern of gene expression elicited by a given toxicant (Nuwaysir et al. 1999). Currently, the application of gene expression technology to ecotoxicology is immature compared to that of toxicology. Reasons for this difference include: (1) the many variables involved in analyzing the status of natural populations, (2) challenges with developing mass produced gene-chips, (3) inherent difficulties in correlating alterations in gene expression with changes at higher levels of biological organization, and (4) issues of variability, uncertainty and interpretation of responses.

There are three main techniques for identifying and characterizing stress-induced genes: polymerase chain reaction (PCR), Northern blot, and microarray analysis. DNA

Table 1 Biomarkers used to evaluate xenobiotic stress focus primarily on aquatic plant species, though some examples for terrestrial plants are also provided. The identity of the xenobiotic, test organism as well as the quantified response is provided for both the biomarker and comparative morphological endpoint (if available). Depending on the metric reported, either percent or fold increases, effective or inhibitory dose or concentration (ED_x, ID_x, EC_x, or IC_x, where x represents the desired level of effect), or lowest observable effect concentrations (LOECs) are given

Biomarker	Xenobiotic/Stress	Organism	Biomarker response	Morphological response	Reference
Gene expression	Copper	*Lemna gibba*	Expression changes at 8–12 μM	Inhibition of growth (50%) at 6 μM	Akhtar et al. (2005)
	Copper	*Chlamydomonas reinhardtii*	Expression changes at 125 μM	Inhibition of growth at 8–125 μM	Jamers et al. (2006)
	Arsenite	*Lemna minor*	Expression changes at 50 μM	NR[f]	Santos et al. (2006)
	Light intensity	*Dunaliella salina*	Expression changes at 1,000 μmol of photons m^{-2} s^{-1}	Reduction in chlorophyll (growth proxy) by tenfold at 1,000 μmol of photons m^{-2} s^{-1}	Park et al. (2006)
Pathway specific metabolites	Atorvastatin	*Lemna gibba*	Reduction in sterol content; EC_{10} = 26.1 μg/L	Inhibition of growth; EC_{10} = 111 μg/L	Brain et al. (2006)
	Lovastatin	*Lemna gibba*	Reduction in sterol content; EC_{10} = 32.8 μg/L	Inhibition of growth; EC_{10} = 133 μg/L	Brain et al. (2006)
	Isoxaflutol	*Cucumis sativus* L	Reduction in carotonoid content by 56% at 5 mM	Inhibition of growth by 39% at 5 mM	Kushwaha and Bhowmik (1999)
	Chlorsulfuron	*Pisum sativum L. var Alaska*	Enzyme inhibition: IC_{50} = 21 nM	Inhibition of root growth by 80% at 28 nM	Ray (1984)
	Rimsulfuron	Russet Burbank potatoes	Increase in 2-aminobutyric acid content by five- to sixfold	NR	Loper et al. (2002)
	Thifensulfuron + Tribenuron	Russet Burbank potatoes	Increase in 2-aminobutyric acid content by threefold	NR	Loper et al. (2002)
	Imazethapyr	Russet Burbank potatoes	Increase in 2-aminobutyric acid content by 30–50%	NR	Loper et al. (2002)

(continued)

Table 1 (continued)

Biomarker	Xenobiotic/Stress	Organism	Biomarker response	Morphological response	Reference
	Glyphosate	*Corydalis sempervirens*	Increase in shikimic acid content up to 100-fold at 10 mM	NR	Amrhein et al. (1983)
	Glyphosate IPA	*Brassica napus L. cv. Iris*	[a]Increase in shikimic acid; ED_{50} = 710 g of a.i. ha^{-1}	Injury visual ranking; ED_{50} = 374 g of a.i. ha^{-1}	Harring et al. (1998)
	Roundup Bio	*Brassica napus L. cv. Iris*	Increase in shikimic acid; ED_{50} = 84 g of a.i. ha^{-1}	Injury visual ranking; ED_{50} = 118 g of a.i. ha^{-1}	Harring et al. (1998)
	Roundup Ultra	*Brassica napus L. cv. Iris*	Increase in shikimic acid; ED_{50} = 164 g of a.i. ha^{-1}	Injury visual ranking; ED_{50} = 190 g of a.i. ha^{-1}	Harring et al. (1998)
	Glyfos	*Brassica napus L. cv. Iris*	Increase in shikimic acid; ED_{50} = 69 g of a.i. ha^{-1}	Injury visual ranking; ED_{50} = 96 g of a.i. ha^{-1}	Harring et al. (1998)
Phase I enzymes	Ethanol	*Helianthus tuberosus* L	Increase in P-450 activity and content by 1.2- and 2.5-fold, respectively at 300 mM	NR	Reichhart et al. (1979)
	Phenobarbital	*Helianthus tuberosus* L	Increase in P-450 activity and content by 1.1-, 1.6-fold, respectively at 4 mM	NR	Reichhart et al. (1979)
	Manganese	*Helianthus tuberosus* L	Increase in P-450 activity and content by 7- and 5.5-fold, respectively at 25 mM	NR	Reichhart et al. (1979)
	Chlorpropham	*Helianthus tuberosus* L	Decrease in P-450 activity and content by 0.6- and 0.6-fold, respectively at 200 μM	NR	Reichhart et al. (1979)
	Monuron	*Helianthus tuberosus* L	Increase in P-450 content by 1.4-fold at 200 μM	NR	Reichhart et al. (1979)
	Dichlobenil	*Helianthus tuberosus* L	Increase in P-450 activity and content by 1.3- and 2-fold, respectively at 200 μM	NR	Reichhart et al. (1979)
	Tributyltin chloride	*Cladophora sp.*	P-450 inhibition by 65-fold at 30 μg/L	NR	Pflugmacher et al. (2000b)
	Tributyltin chloride	*Ceratophyllum demersum*	P-450 inhibition by 2.4-fold at 30 μg/L	NR	Pflugmacher et al. (2000b)

	Tributyltin chloride	*Elodea Canadensis*	P-450 inhibition by 4.4-fold at 30 μg/L	NR	Pflugmacher et al. (2000b)
	2,4,6-trichlorophenol[b]	*Spirodela punctata*	Increase in external peroxidase activity after exposure to 0.1 μM	Inhibition of growth by 10% at 0.3 μM	Jansen et al. (2004)
	2,4,6-trichlorophenol[d]	*Lemna gibba*	Increase in external peroxidase activity after exposure to 0.03 μM	Inhibition of growth by 80% at 0.3 μM	Jansen et al. (2004)
	Anatoxin-a	*Lemna minor*	Increase in peroxidase activity by 50% after 4-d exposure to 25 mg/L	EC_{50} ~5 mg/L for photosynthetic O_2 production	Mitrovic et al. (2004)
	Anatoxin-a	*Chladophora fracta*	Increase in peroxidase activity by 50% after 4-d exposure to 25 mg/L	NR	Mitrovic et al. (2004)
Phase II enzymes	Copper	*Lemna minor*	Increase in glutathione-s-tranferase (GST) activity of 125% at 0.5 μM, reduction to 59% a 10 μM	NR	Teisseire and Guy (2000)
	Lead	*Hydrilla verticillata*	Reduction in glutathione pool by >50% at 2.5 μM	NR	Gupta et al. (1995)
	Paraquat	*Oryza sativa*	Induced GST activity at 100 μM	NR	Zhao and Zhang (2006)
	Lead	*Myriophyllum quitense*	Increase in glutathione reductase, guaiacol peroxidase and cytosolic GST at field sediment concentrations >20 mg/g	NR	Nimptsch et al. (2005)
	Ammonium and nitrate	*Myriophyllum quitense*	Increase in glutathione reductase, guaiacol peroxidase and cytosolic GST at field water concentrations >15 mg/L	NR	Nimptsch et al. (2005)

(continued)

Table 1 (continued)

Biomarker	Xenobiotic/Stress	Organism	Biomarker response	Morphological response	Reference
	PAH's	*Nuphar lutea*	Increase in GST activity at a field sediment concentration of 0.34 mg/kg	NR	Schrenk et al. (1998)
	Microcystin-LR	*Ceratophyllum demersum*	Increase in GST activity after 48-hr exposure to 5 μg/L	NR	Pflugmacher (2004)
	Microcystin-LR	*Phragmites australis*	Increase in GST activity after 24-hr exposure to 0.5 μg/L	NR	Pflugmacher et al. (2001)
	Hexachlorobenzene	*Lemna minor*	Increase in GST activity after 1-, 2- and 7-d exposure to 2 μg/L	NR	Roy et al. (1995)
	Tributyltin chloride	*Cladophora sp.*	Twofold increase in GST activity at 0.3 μg/L, but decrease in GST >12.1 μg/L	NR	Pflugmacher et al. (2000b)
	Tributyltin chloride	*Ceratophyllum demersum*	Threefold increase in GST activity at 0.3 μg/L	NR	Pflugmacher et al. (2000b)
	Tributyltin chloride	*Elodea Canadensis*	One and a half-fold increase in GST activity at 0.3 μg/L	NR	Pflugmacher et al. (2000b)
	Fenchlorim	*Oryza sativa*	Increase in GST activity after a 5-d exposure to 1 μM	NR	Deng and Hatzios (2002)
	Methyl parathion	*Typha latifolia*	Increase in GST activity by 40% after 3-d exposure to 6 mg/L	NR	Amaya-Chávez et al. (2006)
	Anatoxin-a	*Lemna minor*	Increase in GST activity by 40% after 24-hr exposure to 5 and 20 mg/L	5 and 20 mg/L corresponds app. to EC_{50} and EC_{70} for photosynthetic O_2 production	Mitrovic et al. (2004)
	Pyrene	*Chlorella vulgaris*	No change in GST activity at 0.1–1 mg/L	Inhibition of growth; EC_{10} ~1 mg/L after 7-d	Lei et al. (2006)
	Pyrene	*Scenedesmus quadricauda*	Decrease in GST activity by 80% and 30% after 4- and 7-d exposure to 0.6–1.0 mg/L	Inhibition of growth; EC_{70} ~1 mg/L after 7-d	Lei et al. (2006)

	Pyrene	*Scenedesmus platydiscus*	A linear increase in GST activity after 4- and 7-d exposure to 0.1–1.0 mg/L. Fivefold increase at 1.0 mg/L	No effect on growth after 7-d	Lei et al. (2006)
	Pyrene	*Selenastrum cabricornutum*	A linear increase in GST activity after 1-, 4- and 7-d exposure to 0.1–1.0 mg/L. Twofold increase at 1.0 mg/L	No effect on growth after 7-d	Lei et al. (2006)
	Folpet	*Lemna minor*	Increase in GST activity of 50% after 6-hr exposure to 33 μM.	NR	Teisseire and Vernet (2001)
	Propanil	*Lemna minor*	No response in GST activity after 24–168-hr exposure to 0.01–1 mg/L	Inhibition of growth by ~60% at 0.1–1 mg/L	Mitsou et al. (2006)
	Pentachlorophenol	*Eichhornia crassipes*	Increase in GST activity by 50–100% after 24- and 48-hr exposure to 2 μM	NR	Roy and Hanninen (1994)
Phytochelatins	Copper	*Hydrilla verticillata*	Induction of PC2 and PC3 by 5.6-fold and 9.4-fold, respectively, at 1 μM	NR	Srivastava et al. (2006)
	Lead	*Hydrilla verticillata*	Induction of PC2 and PC3 by 1.5-fold and 2.6-fold, respectively, at 10 μM	NR	Gupta et al. (1995)
	Heavy metal mixture	*Stigeoclonium*	Induction of PC levels by twofold	NR	Pawlik-Skowrońska (2001)
	Cadmium	*Lemna aequioctialis*	Increase of PC2 and PC3 to 658 and 118 nM/g FW[c], respectively, from 0 at 20 μM	Inhibition of growth (10.5%) at 20 μM	Yin et al. (2002)
	Cadmium	*Rauvolfia serpentina*	Increase in total PC levels to 20.5 μM/g DW[d] form 0 at 100 μM	NR	Grill et al. (1987)

(continued)

Table 1 (continued)

Biomarker	Xenobiotic/Stress	Organism	Biomarker response	Morphological response	Reference
	Lead	*Rauvolfia serpentina*	Increase in total PC levels to 11.4 µM/g DW form 0 at 1,000 µM	NR	Grill et al. (1987)
	Zinc	*Rauvolfia serpentina*	Increase in total PC levels to 8.5 µM/g DW form 0 at 1,000 µM	NR	Grill et al. (1987)
	Antimony	*Rauvolfia serpentin*	Increase in total PC levels to 8.5 µM/g DW form 0 at 200 µM	NR	Grill et al. (1987)
	Silver	*Rauvolfia serpentina*	Increase in total PC levels to 8.2 µM/g DW form 0 at 50 µM	NR	Grill et al. (1987)
Flavonoids	Copper	*Lemna gibba*	Increase in chalcone synthase activity by 200% at 8 µM	Inhibition of growth by ~30% at 8 µM	Babu et al. (2003)
	Simulated solar radiation (SSR)	*Lemna gibba*	Increase in chalcone synthase activity by 200%	No significant inhibition of growth	Babu et al. (2003)
Stress proteins	Copper	*Enteromorpha intestinalis*	Increase in Hsp70 content by ~55% at 100 µg/L	Inhibition of growth by ~60% at 100 µg/L	Lewis et al. (2001)
	Irgarol	*Enteromorpha intestinalis*	No significant changes in Hsp70 content between 0.1 and 62.5 µg/L	Inhibition of growth by ~75% at 62.5 µg/L	Lewis et al. (2001)
	Zinc	*Raphidocelis subcapitata*	Increase in Hsp70 content; LOEC = 3.6 µM	Cytotoxicity; EC_{50} = 16 µM	Bierkens et al. (1998)
	Selenium	*Raphidocelis subcapitata*	Increase in Hsp70 content; LOEC = 1.6 µM	Cytotoxicity; EC_{50} = 222 µM	Bierkens et al. (1998)

	Sodium dodecyl sulphate	*Raphidocelis subcapitata*	Increase in Hsp70 content; LOEC = 83 μM	Cytotoxicity; EC_{50} = 9.3 μM	Bierkens et al. (1998)
	Carbaryl	*Raphidocelis subcapitata*	Increase in Hsp70 content; LOEC = 10.3 μM	Cytotoxicity; EC_{50} = 25 μM	Bierkens et al. (1998)
	Lindane	*Raphidocelis subcapitata*	Increase in Hsp70 content; LOEC = 9 μM	Cytotoxicity; EC_{50} = 3.1 μM	Bierkens et al. (1998)
	Cadmium	*Fucus serratus*	Increase in Hsp70 content by tenfold at 25 mM	NR	Ireland et al. (2004)
	Cadmium	*Lemna minor*	Increase in Hsp70 content by fivefold at 20 mM	NR	Ireland et al. (2004)
Reactive oxygen species and scavenging enzymes	Copper	*Lemna minor*	Increase in pyrogallol peroxidase activity up to 166% at 10 μM	NR	Teisseire and Guy (2000)
	Copper	*Lemna minor*	Increase in guaiacol peroxidase activity up to 553% at 5 μM	NR	Teisseire and Guy (2000)
	Copper	*Lemna minor*	Increase in catalase activity up to 347% at 10 μM	NR	Teisseire and Guy (2000)
	Copper	*Lemna minor*	Increase in ascorbate peroxidase activity up to 132% at 2 μM, reduction to 72% at 10 μM	NR	Teisseire and Guy (2000)
	Copper + 1,2-dihydroxy-anthraquinone	*Lemna gibba*	Increase in ROS formation (75%) at 4 mM 1,2-dhATQ + 3 mM Cu	Inhibition of growth (70%) at 4 mM 1,2-dhATQ + 3 mM Cu	Babu et al. (2005)
	Copper + 1,2-dihydroxy-anthraquinone	*Lemna gibba*	Increase in superoxide dismutase formation (50–200%) at 4 mM 1,2-dhATQ + 3 mM Cu	Inhibition of growth (70%) at 4 mM 1,2-dhATQ + 3 mM Cu	Babu et al. (2005)

(continued)

Table 1 (continued)

Biomarker	Xenobiotic/Stress	Organism	Biomarker response	Morphological response	Reference
	Folpet	*Lemna minor*	Increase in ascorbate peroxidase and glutathione reductase activity of 155% and 273%, respectively after 24-hr, and 252% increase in catalase activity after 48-hr exposure to 33 μM	NR	Teisseire and Vernet (2001)
	Pentachlorophenol	*Eichhornia crassipes*	Increase in ascorbate peroxidase, gluthatione reductase, and superoxide dismutase activity of 50–100% after 24 and 48-hr of exposure to 2 μM	NR	Roy and Hanninen (1994)
	Copper	*Avicennia marina*	Significant linear relationship between leaf metal accumulation and peroxidase activity	Emergence; LC_{50} = 566 μg/g@ Biomass; EC_{50} = 380 μg/g	MacFarlane and Burchett (2002)
	Lead	*Avicennia marina*	Significant linear relationship between leaf metal accumulation and peroxidase activity	$EC_{50} \geq 800$ μg/g	MacFarlane and Burchett (2002)
	Zinc	*Avicennia marina*	Significant linear relationship between leaf metal accumulation and peroxidase activity	Emergence; LC_{50} = 580 μg/g@ Biomass; EC_{50} = 392 μg/g	MacFarlane and Burchett (2002)
	Zn (Cu and Pb)	*Avicennia marina*	Correlation between leaf Zn (and total Zn, Cu and Pb) content and peroxidase activity in the field.	NR	MacFarlane (2002)
	Copper	*Lemna minor*	Twofold linear increase in catalase activity from 0 to 0.1 mg/L, no change from 0.1 to 0.6 mg/L	Inhibition of growth; EC_{50} = 0.16 mg/L	Teisseire et al. (1998)

	Folpet	*Lemna minor*	Linear increase in catalase activity of 30% from 0 to 10 mg/L, no change from 10 to 40 mg/L	Inhibition of growth $EC_{50} = 7.5$ mg/L	Teisseire et al. (1998)
Pigments	Copper	*Avicennia marina*	Significant linear relationship between leaf accumulated metal and chl-a and -b content.	Emergence; $LC_{50} = 566$ µg/g@ Biomass; $EC_{50} = 380$ µg/g	MacFarlane and Burchett (2002)
	Lead	*Avicennia marina*	No significant linear relationship between leaf accumulated metal and chlorophyll a and b and carotenoid content.	$EC_{50} > 800$ µ/g sediment	MacFarlane and Burchett (2002)
	Zinc	*Avicennia marina*	Significant linear relationship between leaf accumulated metal and chl-a and -b and carotenoid content.	Emergence; $LC_{50} = 580$ µg/g@ Biomass; $EC_{50} = 392$ µg/g	MacFarlane and Burchett (2002)
	Zinc	*Avicennia marina*	Negative correlation between leaf Zn and the chl a/b ration in the field. Not stable temporarily.	NR	MacFarlane (2002)
	Boron	*Oryza sativa*	Reduction in Chl-a content; $EC_{50} = 28.1$, 15.0 and 9.8 mg/L after 1, 2 and 3 week exposure, respectively	Reduction of plant dry weight; $EC_{50} \geq 100$, 33.8 and 15.1 mg/L after 1, 2 and 3 week exposure, respectively	Powell et al. (1996)
	Dichloroacetic acid	*Myriophyllum spicatum*	No significant response for chl-a or -b or carotenoid content	Inhibition of growth, various morphological endpoints; $EC_{50} = 114–340$ mg/L	Hanson et al. (2003)
	Dichloroacetic acid	*Myriophyllum sibiricum*	No significant response for chl-a or -b or carotenoid content	Inhibition of growth, various morphological endpoints; $EC_{50} = 90–400$ mg/L	Hanson et al. (2003)

(continued)

Table 1 (continued)

Biomarker	Xenobiotic/Stress	Organism	Biomarker response	Morphological response	Reference
	Dichloroacetic acid	*Lemna gibba*	No significant response for chl-a or -b or carotenoid content	Inhibition of growth, various morphological endpoints; EC_{50} = 30–180 mg/L	Hanson et al. (2003)
	Acifluorfen	*Lemna minor*	Reduction of chl-a + b and carotenoids; EC_{50} = 290–529 and 287–411 μg/L, respectively	Inhibition of relative growth rates based on leaf area; EC_{50} = 404–475 μg/L	Cedergreen et al. (2007)
	Diquat	*Lemna minor*	Reduction of chl-a + b and carotenoids; EC_{50} = 14–22 and 13–17 μg/L, respectively	Inhibition of relative growth rates based on leaf area; EC_{50} = 20–31 μg/L	Cedergreen et al. (2007)
	Glyphosate	*Lemna minor*	Reduction of chl-a + b and carotenoids; EC_{50} = 18–19 and 18 mg/L, respectively	Inhibition of relative growth rates based on leaf area; EC_{50} = 17–24 mg/L	Cedergreen et al. (2007)
	Mecoprop	*Lemna minor*	Reduction of chl-a + b and carotenoids; EC_{50} = 14–45 and 8–20 mg/L, respectively	Inhibition of relative growth rates based on leaf area; EC_{50} = 8–28 mg/L	Cedergreen et al. (2007)
	Mesotrion	*Lemna minor*	Reduction of chl-a + b and carotenoids; EC_{50} = 6–42 and 6–52 μg/L, respectively	Inhibition of relative growth rates based on leaf area; EC_{50} = 10–23 μg/L	Cedergreen et al. (2007)
	Terbuthylazine	*Lemna minor*	Reduction of chl-a + b and carotenoids; EC_{50} = could not be measured	Inhibition of relative growth rates based on leaf area; EC_{50} = 129–134 μg/L	Cedergreen et al. (2007)
	Copper	*Lemna minor*	Reduction in chl content; EC_{50} of 0.16 mg/L	Inhibition of growth; EC_{50} = 0.16 mg/L	Teisseire et al. (1998)
	Folpet	*Lemna minor*	Reduction in chl content; EC_{50} of 22.8 mg/L	Inhibition of growth; EC_{50} = 7.5 mg/L	Teisseire et al. (1998)
	Cadmium	*Zea mays*	Reduction in chl content at >1.7 μM	Decrease in shoot length and leaf dry weight at >25 μM	Lagriffoul et al. (1998)
	Cadmium	*Helianthus annuus*	Reduction in chl-a and -b; EC_{50} = 181 and 31 μM, respectively	Inhibition of fresh weight and dry weight; EC_{50} = 11 and 33 μM, respectively	Azevedo et al. (2005)

2.4-D	*Myriophyllum aquaticum*	Reduction in chl-a, -b and carotenoid content; EC_{50} = 20, 22 and 19 μg/L, respectively	Inhibition of fresh weight, shoot length and root length; EC_{50} = n.d.[e], n.d. and 50 μg/L, respectively	Turgut and Fomin (2002)
Dichlorprop	*Myriophyllum aquaticum*	Reduction in chl-a, -b and carotenoid content; EC_{50} = 70, 63 and 87 μg/L, respectively	Inhibition of fresh weight, shoot length and root length; EC_{50} = n.d, n.d. and 50 μg/L, respectively	Turgut and Fomin (2002)
Dicamba	*Myriophyllum aquaticum*	Reduction in chl-a, -b and carotenoid content; EC_{50} = 98, 99 and 99 μg/L, respectively	Inhibition of fresh weight, shoot length and root length; EC_{50} = n.d, n.d. and 106 μg/L, respectively	Turgut and Fomin (2002)
Pyridate	*Myriophyllum aquaticum*	Reduction in chl-a, -b and carotenoid content; EC_{50} = 636, 555 and 653 μg/L, respectively	Inhibition of fresh weight, shoot length and root length; EC_{50} = n.d, n.d. and 100 μg/L, respectively	Turgut and Fomin (2002)
Propiquizafop	*Myriophyllum aquaticum*	Reduction in chl-a, -b and carotenoid content; EC_{50} = 1.1, 1.2 and 1.4 mg/L, respectively	Inhibition of fresh weight, shoot length and root length; EC_{50} = 1.7, n.d. and 0.5 mg/L, respectively	Turgut and Fomin (2002)
Terbuthryn	*Myriophyllum aquaticum*	Reduction in chl-a, -b and carotenoid content; EC_{50} = 67, 69 and 68 mg/L, respectively	Inhibition of fresh weight, shoot length and root length; EC_{50} = not detectable	Turgut and Fomin (2002)
Triflusulfuron methyl	*Myriophyllum aquaticum*	Reduction in chl-a, -b and carotenoid content; EC_{50} = 37, 36 and 45 μg/L, respectively	Inhibition of fresh weight, shoot length and root length; EC_{50} = 30, 18 and 65 μg/L, respectively	Turgut and Fomin (2002)
Thifensulfuron methyl	*Myriophyllum aquaticum*	Reduction in chl-a, -b and carotenoid content; EC_{50} = 9, 11 and 12 μg/L, respectively	Inhibition of fresh weight, shoot length and root length; EC_{50} = 3, 6 and 11 μg/L, respectively	Turgut and Fomin (2002)
Metsulfuron methyl	*Myriophyllum aquaticum*	Reduction in chl-a, -b and carotenoid content; EC_{50} = 62, 0.88 and 0.88 μg/L, respectively	Inhibition of fresh weight, shoot length and root length; EC_{50} = n.d., 7 and 4 μg/L, respectively	Turgut and Fomin (2002)

(continued)

Table 1 (continued)

Biomarker	Xenobiotic/Stress	Organism	Biomarker response	Morphological response	Reference
	Amidosulfuron	*Myriophyllum aquaticum*	Reduction in chl-a, -b and carotenoid content; EC_{50} = 325, 325 and 325 μg/L, respectively	Inhibition of fresh weight, shoot length and root length; EC_{50} = 970, 974 and n.d. μg/L, respectively	Turgut and Fomin (2002)
	Rimsulfuron	*Myriophyllum aquaticum*	Reduction in chl-a, -b and carotenoid content; EC_{50} = 149, 149 and 149 μg/L, respectively	Inhibition of fresh weight, shoot length and root length; EC_{50} = 124, 97 and 1498 μg/L, respectively	Turgut and Fomin (2002)
	Glyphosate	*Myriophyllum aquaticum*	Reduction in chl-a, -b and carotenoid content; EC_{50} = 222, 222 and 222 μg/L, respectively	Inhibition of fresh weight, shoot length and root length; EC_{50} = 1999, 2040 and 1998 μg/L, respectively	Turgut and Fomin (2002)
	Trifluralin	*Myriophyllum aquaticum*	Reduction in chl-a, -b and carotenoid content; EC_{50} = 323, 353 and 333 μg/L, respectively	Inhibition of fresh weight, shoot length and root length; EC_{50} = 279, 338 and 418 μg/L, respectively	Turgut and Fomin (2002)
	Pendimethalin	*Myriophyllum aquaticum*	Reduction in chl-a, -b and carotenoid content; EC_{50} = 14, 16 and 16 mg/L, respectively	Inhibition of fresh weight, shoot length and root length; EC_{50} = n.d., 10.7 and n.d. mg/L, respectively	Turgut and Fomin 2002)
	Chlorthalonil	*Myriophyllum aquaticum*	Reduction in chl-a, -b and carotenoid content; EC_{50} = 2.7, 2.5 and 2.6 mg/L, respectively	Inhibition of fresh weight, shoot length and root length; EC_{50} = n.d, 2.9 and 0.5 mg/L, respectively	Turgut and Fomin (2002)
	Propiconazole	*Myriophyllum aquaticum*	Reduction in chl-a, -b and carotenoid content; EC_{50} = 3.9, 3.9 and 3.8 mg/L, respectively	Inhibition of fresh weight, shoot length and root length, EC_{50} = 1.3, 0.5 and n.d. mg/L, respectively	Turgut and Fomin (2002)
	Parathion	*Myriophyllum aquaticum*	Reduction in chl-a, -b and carotenoid content; EC_{50} = 16, 16 and 115 mg/L, respectively	Inhibition of fresh weight, shoot length and root length; EC_{50} = 8.8, 6.9 and 16.3 mg/L, respectively	Turgut and Fomin (2002)

Metsulfuron methyl	*Myriophyllum aquaticum*	Reduction in chl-a, -b and carotenoid content; EC_{50} = 1.6, 2.3 and 2.3 nM, respectively	Inhibition of shoot and root length; EC50 = 18 and 10 nM, respectively	Turgut et al. (2003)
Thifensulfuron methyl	*Myriophyllum aquaticum*	Reduction in chl-a, -b and carotenoid content; EC_{50} = 22, 28 and 31 nM, respectively	Inhibition of shoot and root length; EC50 = 15 and 28 nM, respectively	Turgut et al. (2003)
Chlorsulfuron	*Myriophyllum aquaticum*	Reduction in chl-a, -b and carotenoid content; EC_{50} = 1.6, 2.8 and 2.8 nM, respectively	Inhibition of shoot and root length; EC50 = 12 and n.d. nM, respectively	Turgut et al. (2003)
Tribenuron	*Myriophyllum aquaticum*	Reduction in chl-a, -b and carotenoid content; EC_{50} = 3.1, 16 and 13 µM, respectively	Inhibition of shoot and root length; EC50 = 0.5 and 2.9 µM, respectively	Turgut et al. (2003)
Paraquat	*Lemna minor*	Decrease in chl-a, -b and carotenoid content to ~23%, 17% and 4% of control at 100 µg/L	Inhibition of growth rate; EC_{50} ~100 µg/L	Frankart et al. (2003)
Norflurazon	*Lemna minor*	Decrease in chl-a, -b and carotenoid content to ~73%, 62% and 62% of control at 100 µg/L, respectively	Inhibition of growth rate by 15% at 1–100 µg/L	Frankart et al. (2003)
Flazasulfuron	*Lemna minor*	Decrease in chl-a, -b and carotenoid content to ~85%, 80% and 90% of control at 100 µg/L, respectively	Inhibition of growth rate by 25% at 100 µg/L	Frankart et al. (2003)
Atrazine	*Lemna minor*	Decrease in chl-a, -b and carotenoid content to ~79%, 71% and 74% of control at 100 µg/L, respectively	Inhibition of growth by rate 20% at 100 µg/L	Frankart et al. (2003)
Creosote	*Lemna gibba*	Reduction in chl content after 8 d; EC50=3.4 mg/L	Inhibition of growth; EC_{50}=7.2 mg/L	(Marwood et al. 2001)
Creosote	*Myriophyllum spicatum*	Reduction in chl content after 12 d; EC50 = 5.6 mg/L	Inhibition of growth; EC_{50} = 2.6 mg/L	Marwood et al. (2001)

(continued)

Table 1 (continued)

Biomarker	Xenobiotic/Stress	Organism	Biomarker response	Morphological response	Reference
	Chromium	*Spirodela polyrhiza*	Reduction in chl-a and carotenoid content; EC50 ≅40 and 75 μM	Inhibition of relative growth rate; EC50 ≅140 μM	Appenroth et al. (2001)
Photosynthesis	Paraquat	*Lemna minor*	Inhibition of F_v/F_m, Φ_{PSII} and F_v'/F_m' by 14%, 52% and 7%, respectively at 10 μg/L	Inhibition of growth by 13% at 10 μg/L	Frankart et al. (2003)
	Norflurazon	*Lemna minor*	Inhibition of F_v/F_m, Φ_{PSII} and F_v'/F_m' by 31%, 52% and 37%, respectively at 10 μg/L	Inhibition of growth by 14% at 100 μg/L	Frankart et al. (2003)
	Atrazine	*Lemna minor*	Inhibition of F_v/F_m, Φ_{PSII} and F_v'/F_m' by 11%, 78% and 34%, respectively at 10 μg/L	Inhibition of growth by 18% at 100 μg/L	Frankart et al. (2003)
	Flazasulfuron	*Lemna minor*	Inhibition of F_v/F_m, Φ_{PSII} and F_v'/F_m' by 6%, 9% and 4%, respectively at 10 μg/L	Inhibition of growth by 26% at 100 μg/L	Frankart et al. (2003)
	Irgarol	*Enteromorpha intestinalis*	Inhibition of F_v/F_m by ~75% at 62.5 μg/L	Inhibition of growth by ~75% at 62.5 μg/L	Lewis et al. (2001)
	Cd	*Helianthus annuus*	Inhibition of F_v/F_m; EC_{50} = 296 μM	Inhibition of fresh and dry weight; EC_{50} = 11 and 33 μM	Azevedo et al. (2005)
	Creosote	*Myriophyllum spicatum*	Inhibition of F_v/F_m and $\Delta F/F_m$ EC_{50} = 0.28 and 0.30 mg/L, respectively, for single exposure treatment	Seasonal biomass accumulation was inhibited to a similar extent as fluorescence after 8 d	Marwood et al. (2003)
	Creosote	*Lemna gibba*	Inhibition of F_v/F_m and $\Delta F/F_m$; EC_{50} = 36 and 13 mg/L, respectively	EC_{50} for growth was 7.2 mg/L	Marwood et al. (2001)

Creosote	*Myriophyllum spicatum*	Inhibition of F_v/F_m and $\Delta F/F_m$; $EC_{50} = 13$ and 15 mg/L, respectively	EC_{50} for growth was 2.6 mg/L	Marwood et al. (2001)
Sewage treatment hydrophobic component	*Closterium ehrenbergii*	Q_N decreased to zero after 2 hr and C.A./F_{vp} increased two-fold after 96-hr exposure to 1.5 increase of hydrophobic component	Inhibition of growth by 30% at a 1.5-fold increase of hydrophobic component	Juneau et al. (2003)
Sewage treatment hydrophobic component	*Lemna gibba*	Q_N reduced to 50% and C.A./F_{vp} reduced to 70% of control after 96 hr of exposure to a 40-fold increase of hydrophobic component	NR	Juneau et al. (2003)
Chromium	*Spirodela polyrhiza*	Inhibition of photosynthetic O_2 evolution; $EC_{50} \cong 100$ μM	Inhibition of relative growth rate; $EC_{50} \cong 140$ μM	Appenroth et al. (2001)
Chromium	*Spirodela polyrhiza*	Eighteen fluorescence parameters, directly measured or derived, were either more, or as sensitive as growth measures	Inhibition of relative growth rate; $EC_{50} \cong 140$ μM	Appenroth et al. (2001)
Linuron	*Lemna minor*	Inhibition of $\Delta F/F'_m$ within 90 min after exposure to >160 nM	Inhibition of accumulated biomass; $EC_{50} \cong 40$ nM	Hulsen et al. (2002)
Copper	*Lemna minor*	Inhibition of ΔDFI and F_v/F_m; $EC_{50} = 2.2$ and >10 mg/L after 24 hr, respectively	Inhibition of growth $EC_{50} \sim 0.1$ mg/L	Drinovec et al. (2004)
Cadmium	*Lemna minor*	Inhibition of ΔDFI and F_v/F_m; $EC_{50} = 2.2$ and >500 mg/L after 24 hr, respectively	NR	Drinovec et al. (2004)

(continued)

Table 1 (continued)

Biomarker	Xenobiotic/Stress	Organism	Biomarker response	Morphological response	Reference
	Zinc	*Lemna minor*	Inhibition of ΔDFI and F_v/F_m; $EC_{50} = 2.2$ and 2,234 mg/L after 24 hr, respectively	NR	Drinovec et al. (2004)
	Irgarol	*Zostera marina*	Inhibition of F_v/F_m after 10 d; LOEC = 0.5 and $EC_{50} = 1.1$ μg/L	Inhibition of leaf specific biomass; LOEC = 1.0 μg/L	Chesworth et al. (2004)
	Diuron	*Zostera marina*	Inhibition of F_v/F_m after 10-d exposure; LOEC = 1.0 and $EC_{50} = 3.2$ μg/L	Inhibition of leaf specific biomass; LOEC = 5.0 μg/L	Chesworth et al. (2004)

[a]Concentration expressed in terms of active ingredient (a.i.)
[b]Similar results for the halogenated phenols 2,4,6-tribromophenol and 2,4,6-trifluorophenol
[c]Expressed in units per g fresh weight
[d]Expressed in units per g dry weight
[e]N.d. is "not determinable". Tested concentration range is not given
[f]*NR* not reported

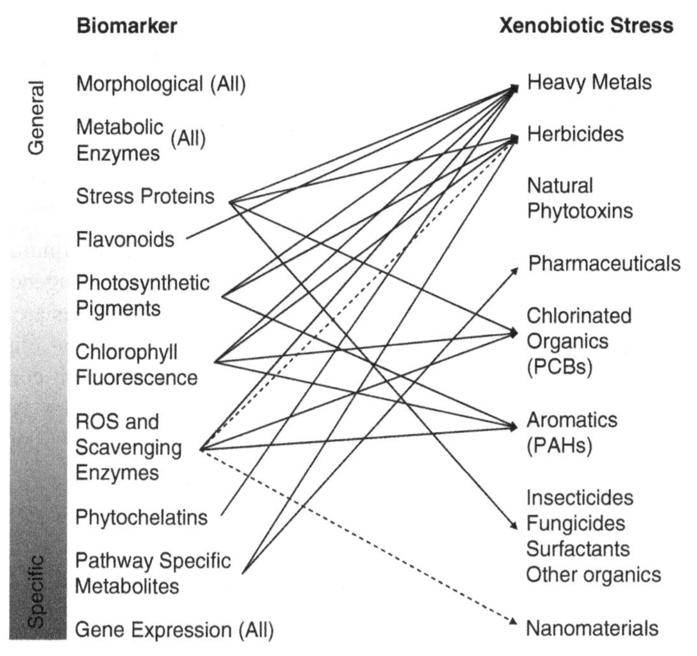

Fig. 1 Dendrogram indicating which biomarkers are appropriate for which xenobiotic stress types. Solid arrows indicate a strong, and dotted arrows a weak association. This diagram does not represent all applications of any biomarker, rather it shows the groups of contaminants that have been evaluated by a given biomarker. (All) denotes that the biomarker is potentially appropriate for all types of stressors (the associated arrows have been excluded to increase readability for other biomarker/stress associations)

microarrays facilitate genetic profiling of responses to environmental stress, although only a few complete genomes have been sequenced for organisms typically employed in ecotoxicological evaluations. For incomplete genomes, heterologous hybridizations across strains and closely related species are possible for a given gene, provided the sequence divergence is limited (Lettieri 2006). However, despite the advent of PCR techniques and gene array analysis, northern hybridization (an electrophoresis technique) using cDNA probes still remains the dominant method for detection and quantification of mRNA levels. There are limited examples that employ these gene expression techniques for evaluating the stress effects of heavy metal and photosynthetic-disrupting contaminants in aquatic higher plants (Akhtar et al. 2005; Jamers et al. 2006; Park et al. 2006; Sánchez-Estudillo et al. 2006; Santos et al. 2006).

2.2 *Quantification*

The isolation of RNA, amplification and cloning sequence analysis of cDNAs, and RNA quantification are facilitated mainly by using specialized kits (Akhtar et al. 2005; Jamers et al. 2006; Santos et al. 2006). Identification and gene amplification typically employ PCR assays, where changes in cDNA expression levels can be quantified (Akhtar et al. 2005; Jamers et al. 2006; Santos et al. 2006). In cloning and sequence analysis, purified bands on PCR gels are typically identified by comparing them to cDNA sequences in electronic gene databases (Akhtar et al. 2005). Northern blot analysis remains the preferred method to identify RNA sequences known to change in response to toxicant stress. Isolated RNA, separated by electrophoresis, and cDNAs isolated from PCR labeled with [^{32}P]dCTP, are used as probes and quantified by autoradiography (Akhtar et al. 2005).

DNA arrays (gene chips) are used to profile gene expression by simultaneously monitoring expression levels of thousands of genes. Using high-stringency conditions for hybridization, both qualitative and quantitative measurements are possible using fluorophore-based detection. Microarrays, based on cDNAs, or oligonucleotides are hybridized with fluorescent cDNA probes generated from control and test RNA products; these form labeled products with different fluors. A ratio of fluors other than one, indicates differentially expressed genes between two populations (Jamers et al. 2006; Nuwaysir et al. 1999).

2.3 Applications to Evaluate Xenobiotic Stress

Studies employing gene expression techniques to evaluate xenobiotic and physical stress in aquatic plants have shown variable results. Akhtar et al. (2005) used differential display PCR (ddPCR) and subsequent northern hybridization, in *L. gibba* plants exposed to copper, to detect changes in cDNA levels for genes encoding proteins known to increase in response to environmental and xenobiotic stress. Levels of the larger transcript of the heme-activated protein (HAP) transcription factor complex (regulated by light and suppressed by oxidative stress) decreased upon exposure to copper (Akhtar et al. 2005). Copper is known to promote the formation of ROS, which disrupt biological membranes (see Sect. 9). Similarly, cDNAs also showed marked decreases for the chloroplast nucleoid DNA-binding protein. This protein is probably involved in chloroplast gene expression, where down-regulation may allow plants to synthesize more D1 protein (affected by copper), thereby counteracting the effect of copper-induced damage to photosystem II (PSII) (Akhtar et al. 2005). In contrast, mRNA transcripts for callose synthase, serine decarboxylase (SDC), acetyl–coenzyme-A carboxylase (ACCase), and a heat shock protein (HSP: HSP90) showed increased transcription abundance in exposed plants (Akhtar et al. 2005). Callose synthase, which controls the deposition of callose into cell walls, is induced by wounding, pathogen challenge, and metal stress. SDC, which controls the flux of ethanolamine into the polar head groups of phospholipids, and ACCase, which contributes solely to the production of fatty acids were probably induced in response to ROS formation (Akhtar et al. 2005). HSPs are known to be induced by heavy metals, and may respond to other types of environmental stress. Hence, the changes in gene expression may potentially be explained by the physiological effects of Cu, in this case.

Santos et al. (2006) evaluated expression levels of components and intermediates of the ubiquitin (Ub)/proteasome pathway in *Lemna minor* exposed to arsenite, using relative quantitative reverse transcription PCR (RT–PCR). This pathway was chosen because the Ub/proteasome system is a tightly regulated and highly specific for targeted proteolysis, and arsenite is known to promote ROS formation. At non-lethal arsenite concentrations (50 μM $NaAsO_2$), a moderate but significant increase in the steady-state abundance of transcripts was observed for each gene, suggesting ROS-mediated induction of the Ub/proteasome system (Santos et al. 2006).

Park et al. (2006) evaluated the up-regulation of photoprotection and PSII-repair gene expression in *D. salina,* when this species was exposed to irradiation. Approximately 1% of 1,038 expressed sequence tags (ESTs) investigated were assigned to genes that coded for known light-stress-inducible proteins. From these, four genes encoding the putative carotinoid biosynthesis related gene (Cbr), Fe-superoxide dismutase (Fe-SOD), ATP-dependent endopetidase (Clp), and ascorbate-peroxidase APX gene transcripts were evaluated for differential expression under conditions of low and high light (Park et al. 2006). When probed by northern blot analysis, the APX, Cbr, Fe-SOD, and Clp genes, gave increased levels of expression (3.2-, 5.4-, 4.8- and 2.6-fold, respectively) after 48 hr of exposure, when compared with controls. Under similar conditions, Jin et al. (2003) found that exposure of *D. salina* to high light only induced a 10% reduction in chlorophyll levels and a 50% reduction in carotinoid levels.

Jamers et al. (2006) evaluated the effect of copper exposure on gene expression profiles in the green algae *Chlamydomonas reinhardti,* using microarray analysis. Of the 2,407 differentially ESTs, 362 sequences were identified, based on comparison to genome databases. Several genes that involve oxidative stress defense mechanisms, including glutathione peroxidase and a probable glutathione *S*-transferase sequence, showed increased expression by as much as 20 times (Jamers et al. 2006). Similarly, transcription of lipoxygenase, a key enzyme in the execution of apoptosis induced by oxidative stress, increased 7.2-fold, and several genes concerned with protein damage (recognition, repair and/or proteolysis) were also differentially expressed (Jamers et al. 2006). Inexplicably, genes encoding photosynthetic electron transport carriers (cytochrome b6/f and ferredoxin) were down regulated when compared to genes involved in mitochondrial electron transport. Furthermore, transcription levels coding for a putative mitochondrial uncoupling protein involved in antioxidant defense decreased at all time points (Jamers et al. 2006). Several other genes also displayed differential responses, including a number evaluated under copper deficient conditions. Although gene expression appears to show comparable sensitivity to that of algal growth rate, which decreased continuously at concentrations above 8 μM (25, 55 and 125 μM), microarray responses were not always intuitive, expected or logical.

2.4 Advantages

Gene expression assays provide a means to simultaneously assess the expression of thousands of genes at the mRNA level. Therefore, these assays offer a sophisticated, high throughput screening tool for possible identification of molecular mechanisms of, and novel biomarkers for, environmental stresses (Jamers et al. 2006). In addition, measurement of gene expression levels, after chemical exposure, may facilitate the formation of a "genetic signature" from the pattern of gene expression changes (Park et al. 2006). Gene expression techniques thus provide a snapshot of the repertoire of genes expressed by a cell or tissue at the time of harvest and RNA purification (Nygaard and Hovig 2006). With the advent of amplification techniques, only a small amount of isolated RNA is required for these analyses.

2.5 Disadvantages

A major challenge lies in taking into account intrinsic sources of variability in gene expression levels including: different physiologic states, age, cell types, time after exposure, environmental conditions and genetic polymorphisms in natural populations (Lettieri 2006). Moreover, the expression of certain genes may vary considerably, even under tightly controlled experimental conditions (Lettieri 2006). Another challenge is the inherent difficulty in relating alterations in gene expression with, and consequences of, changes higher up the biological continuum, such as morphological, population or ecosystem effects. Furthermore, the relative sensitivity of gene expression techniques, compared to morphological endpoints is still uncertain. Genomics is currently limited by a series of factors: there are few species for which microarray gene-chips are available; the technology is expensive, which limits repeat testing (Lettieri 2006); gene expression assays are laborious and time consuming and many toxicant-stress genes are similar to those induced by environmental stress, rendering only lab-experiments valid. Moreover, toxicant-induced genes are often similar across a wide range of toxicants, making gene expression less certain as a "mode-of-action" diagnostic tool than initially proposed (Baerson et al. 2005). In PCR analyses, there is concern that more than one cDNA can be obtained from a single PCR band, which is attributed to nearly identically sized cDNA fragments that co-migrate on gels (Akhtar et al. 2005). Changes in transcript abundance identified by PCR analysis may constitute false positives, when compared with northern hybridization analysis; thus, although PCR is a powerful technique, it must be used with caution (Akhtar et al. 2005).

3 Pathway Specific Metabolites

3.1 Overview

Because different xenobiotic stresses may produce similar metabolic reactions, analysis of changes in metabolites from disrupted biochemical pathways can provide more specific information regarding a toxicant's nature, and provide stronger correlation between the measured biomarker and the biological effect (Ernst and Peterson 1994). When evaluating contaminant effects on metabolic pathways, measuring the activity of a target enzyme directly, or through variations in metabolite concentrations (either upstream or downstream) provides the greatest diagnostic resolution. Furthermore, changes in metabolite concentrations provide the foundation on which all other stessor-induced effects depend. In plants, herbicide studies best illustrate the utility of metabolites or enzymes as biomarkers for displaying specificity, sensitivity, and relevance to morphological effects. Similar examples exist for pharmaceuticals, another group of highly specific biologically active compounds.

3.2 Quantification

Procedures for extraction and analysis vary markedly, and depend on the physiochemical properties of the metabolite(s) of interest. For tests with *L. gibba*, extraction, purification and quantification procedures exist for polar and non-polar compounds (plastoquinone, ubiquinone, β-sitosterol and stigmasterol) (Brain et al. 2006), and are based on protocols for algae (Schwender et al. 1996) and seaweeds (Sánchez-Machado et al. 2004). Extraction and quantification methods also exist for metabolites that accumulate in the presence of ALS- (acetolactate synthase) inhibiting herbicides (2-aminobutyric acid) (Loper et al. 2002) and glyphosate (shikimic acid), respectively (Harring et al. 1998; Lydon and Duke 1988). Published methods for the extraction and purification of enzymes (ALS and 5-enolpyruvylshikimic acid-3-phosphate (EPSP)-synthase) are also available (Ray 1984; Amrhein et al. 1983).

3.3 Applications to Evaluate Xenobiotic Stress

Although biologically active compounds target many plant pathways, few studies demonstrate and compare fluctuations in metabolite production to growth endpoints. Brain et al. (2006) evaluated the pathway-specific effects of statin class blood lipid regulators in *L. gibba* plants. Statins are competitive inhibitors of 3-hydroxy-3-methylglutaryl coenzyme-A reductase (HMGR), the rate determining enzyme of the mevalonic acid (MVA) pathway, responsible for synthesizing a myriad of compounds, including sterols. HPLC-UV (ultraviolet) analysis of plant extracts showed significantly decreased concentrations of both stigmasterol and β-sitosterol, which are critical components of plant membranes that regulate morphogenesis and development (Brain et al. 2006). The statin-mediated reductions of sterol concentrations, critical to plant function, accounted for the resulting phytotoxicity and growth inhibition. Furthermore, EC_{10} values for atorvastatin and lovastatin were 2–3 times more sensitive effect-measures than were fresh-weight measures (Brain et al. 2006).

Other examples of pathway specific responses can be drawn from herbicide mode-of-action studies. Kushwaha and Bhowmik (1999) found the treatment of cucumber cotyledons with isoxaflutole (isoxazole class of herbicides), which inhibits the enzyme *p*-hydroxyphenyl pyruvate dioxygenase (HPPD), significantly inhibited carotinoid biosynthesis. HPPD is responsible for converting *p*-hydroxyphenyl pyruvate to homogentisate, a precursor to plastoquinone, which is required for carotinoid biosynthesis; carotinoids are critical protective pigments. Carotinoid content was 30% more sensitive as an indicator of effect than was measuring fresh weight (Kushwaha and Bhowmik 1999). However, greater specificity would have been provided by using plastoquinone production as the key determinate. This is because several herbicides and other abiotic stresses (changes in irradiance, CO_2 and nutrient availability) can cause reductions in carotinoid levels.

Evaluating metabolite production and enzyme activity has also been useful when assessing modes of action, specificity and sensitivity of amino acid synthesis-inhibiting herbicides, including sulfonylureas and glyphosate. Sulfonylurea herbicides interfere with the production of branched-chain amino acids by inhibiting ALS (Ray 1984). In experiments with excised pea (*Pisum sativum* L. var Alaska) roots, Ray (1984) found that ALS activity was specifically inhibited by exposure to chlorsulfuron at concentrations which also caused significant inhibition of root growth.

Although not empirically compared to a morphological parameter such as growth, Loper et al. (2002) demonstrated the utility of 2-aminobutyric acid (2-aba) as an exclusive biomarker for ALS-inhibiting herbicides. Present at ppm (parts per million) concentrations in plant tissue, 2-aba is a precursor for branched chain amino acid synthesis utilized by ALS, and is therefore a useful marker. Analytically, macro-changes in 2-aba content are easy to detect (i.e., 1 μg/g control vs 5–15 μg/g treated) compared to analyzing matrix (water) herbicide concentrations (Loper et al. 2002). Furthermore, the use of a single chemical marker method, specific to ALS-inhibiting herbicides, is simpler and less expensive (Loper et al. 2002). Amounts of 2-aba, up to 6 times that of controls, were detected for several sulfonylurea and imidazole herbicides, and therefore constituted a highly sensitive marker method. Because 2-aba begins to build up in susceptible plants within hour, exposure to ALS-inhibitor herbicides can be assessed prior to symptom expression, which usually takes d or week to develop (Loper et al. 2002).

Similar results were obtained for glyphosate, which inhibits EPSP, an enzyme critical in the shikimic acid pathway responsible for production of aromatic amino acids (phenylalanine, tryptophan and tyrosine) (Amrhein et al. 1980). Blockage of the pathway results in accumulation of high levels of shikimic acid and, to a lesser extent, shikimic acid-3-phosphate (metabolic precursors) and loss of feedback control (Amrhein et al. 1980; Jensen 1985). Exposure to glyphosate concentrations of 0.5, 1, 2, 3, 5, and 10 mM *in vivo* caused respective increases in shikimic acid concentration of 44, 49, 56, 56, 87, and 100 times that of controls in cultured plant cells (*Corydalis sempervirens*) (Amrhein et al. 1980). These results further demonstrate the utility of analyzing pathway-specific metabolites as biomarkers of exposure and effects.

Harring et al. (1998) found that the accumulation of shikimic acid in leaves of 3-week-old oil seed rape (*Brassica napus* L. cv. Iris) was clearly dependent on the treatment concentration of glyphosate. However, the accumulation of shikimic acid was not elevated by adding glyphosate in excess of 200 g of active ingredient (a.i.) ha^{-1}; this probably results from Michaelis-Menten saturation of EPSP synthase

(Harring et al. 1998). With the exception of the pure isopropylamine salt of glyphosate (IPA), measuring shikimic acid accumulation 48 hr after exposure was more sensitive than visually assessing plant injury 14 d after exposure (Harring et al. 1998). The ranking of efficacy was the same for shikimic acid accumulation and visual assessment; this supports the hypothesis that measuring shikimic acid is a viable, sensitive and more rapid alternative to the visual assessment approach (Harring et al. 1998).

3.4 Advantages

In plants, toxicant-induced effects are first manifested at the biochemical level. Therefore, measuring metabolite fluctuations not only provides information on mechanism of action, but often also provides greater sensitivity and more rapid evidence of stress than does growth endpoints. Effects at the biochemical level typically occur within minutes to hour, whereas, effects at higher levels of biological organization, particularly the morphological level, may take several d to week. In terms of diagnostic resolution, evaluations of pathway specific metabolites provide the most specific response for a given stressor or group of stressors, and, in contrast to many other biomarkers, these are suitable for use under field conditions, because they are not confounded by different forms of environmental stress (Loper et al. 2002). If the biochemical effect can be related to effects at the physiological or morphological level, pathway specific metabolites can be used as sub-lethal indicators for risk assessment.

3.5 Disadvantages

Measurement of metabolites requires an in-depth understanding of the target pathway's biochemistry. Moreover, measuring metabolites or enzymes requires robust analytical capacity, and methods vary substantially for a given compound. Hence, specific methods typically must be developed for the target organism(s) and metabolite(s). The analytical rigor associated with pathway-specific metabolite investigations often require sophisticated equipment, and are expensive.

4 Metabolic Enzymes: Phase I Metabolic Enzymes

4.1 Overview

Plants are equipped with a remarkably versatile system that protects them from the potentially phytotoxic actions of xenobiotics (Kreuz et al. 1996). A multitude of plant enzymes, primarily cytochrome P-450s or monooxygenases, metabolize drugs, herbicides and other organic xenobiotics to non-phytotoxic products. P-450s are microsomal heme-thiolate proteins similar to those characterized in the endoplasmic reticulum of mammalian livers (Kreuz et al. 1996; Paquette et al. 2000). P-450s typically catalyze the primary step in the detoxification process, a metabolic attack through oxidation or hydrolysis, which usually serves to introduce a hydroxyl, amino or carboxylic acid functional group (Kreuz et al. 1996). The molecule then enters phase II detoxification, which involves conjugation to sugars, amino acids, thiols, etc., via enzymes such as glucosyltransferases and glutathione *S*-transferases, which catalyze the more common conjugations. The resulting conjugates are generally inactive at the initial target site, are more hydrophilic and less mobile in the plant than the parent toxicant, and are susceptible to further processing. Such processing may entail secondary conjugation, degradation and/or compartmentation. Phase II enzymes will be reviewed in the following Sect. 5.

In mammalian systems, Phase I enzymes are induced by a myriad of compounds: polychlorinated dibenzo-p-dioxins (PCDDs) and dibenzofurans (PCDFs), polychlorinated biphenyls (PCBs), polycyclic aromatic hydrocarbons (PAHs), other halogenated biphenyls, terphenyls, pesticides, metals and natural or biogenic substances (Whyte et al. 2000). Similar classes of contaminants induce phase I enzymes in plant systems; for example, tributyltin chloride (Pflugmacher et al. 2000b), manganese, ethanol, phenobarbital and the herbicides dichlobenil and monuron (Reichhart et al. 1980).

4.2 Quantification

The EROD (ethoxyresorufin-O-deethlyase) assay is the most commonly used assay to measure cytochrome P-450 induction (Whyte et al. 2000). EROD activity describes the rate of the CYP1A-mediated deethylation of a phenoxazone ether substrate, 7-ethoxyresorufin (7-ER) in the microsomal fraction, leading to the formation of a hydroxylated product resorufin (Whyte et al. 2000). CYP1A is a gene that encodes for a member of the cytochrome P-450 superfamily of enzymes. Catalytic activity toward 7-ER is a functional indicator of the amount of CYP1A present in a tissue sample measured fluorometrically as resorufin, at 530 and 580 nm (Pflugmacher et al. 2000b; Whyte et al. 2000). Activity in the sample is standardized based on protein content of the microsomes (Bradford 1976; Whyte et al. 2000). Plant tissue samples for microsomal and soluble protein extracts can be prepared as described by Pflugmacher et al. (2000b).

P-450 activity has also been measured using a cinnamic acid hydroxylation assay, in which production of p-coumaric acid from trans-[3-^{14}C]cinnamic acid is measured by radiochromatography and scintillation counting (Reichhart et al. 1980). The microsomal preparation for this assay is similar to that employed for the EROD assay. Total P-450 protein content can be quantified spectrophotometrically (Reichhart et al. 1980).

4.3 Applications to Evaluate Xenobiotic Stress

Investigations of P-450 content and enzyme activity in plants exposed to contaminants are sparse, although examples exist which demonstrate the utility of these metabolic enzymes as biomarkers. Reichhart et al. (1979) assessed the effect of manganese, iron, ethanol, phenobarbital and several herbicides on cytochrome P-450 (trans-cinnamic acid 4-hydroxylase) activity and content in a wound-induced tuber tissue system from Jerusalem artichoke (*Helianthus tuberosus* L.). At high concentrations, ethanol, isopropanol and methanol increased P-450 content and activity, although butanol was found to be highly inhibitory. Phenobarbital was a weak inducer of P-450, whereas manganese exposure substantially increased P-450 content and activity (Reichhart et al. 1979). Exposure to several herbicides produced variable responses; chlorpropham (mitosis inhibitor) inhibited P-450 induction and activity, whereas dichlobenil (cellulose biosynthetic inhibitor) and monuron (PSII inhibitor) increased P-450 content (Reichhart et al. 1979). No gross morphological or physiological parameters were reported as a basis of comparison.

Pflugmacher et al. (2000b) found that tributyltin chloride (TBTCl) significantly decreased cytochrome P-450 activity in a concentration-dependent manner using the EROD assay in *Cladophora sp.* microsomal fractions. *Ceratophyllum demersum* and *Elodea canadensis* also experienced significantly reduced enzyme activity when exposed to TBTCl (Pflugmacher et al. 2000b). The inhibition of P-450 activity was consistent with results found in freshwater fish, e.g., rainbow trout (*Oncorhynchus mykiss*), European eel (*Anguilla anguilla*), or bullhead (*Cottus gobio*). Similar to aforegoing results of Reichhart et al. (1979), no corresponding gross morphological measurements were reported. Therefore, the relative sensitivities of the P-450 activity assays cannot be compared. Moreover, because a comprehensive evaluation of P-450 activity and content assays across an array of chemicals in plants is lacking, it is impossible to compare the specificity of these assays. However, in fish, a large number of compounds induce EROD activity, while others are inhibitors (Whyte et al. 2000). In plants, EROD (CYP1A1) activity also varies with tissue/compartment (roots, stems and leaves), and is reported to be higher in leaves than roots by a factor of two in *Phragmites australis* (Pflugmacher et al. 1999).

4.4 Advantages

Assays and protocols for rapid analysis have been developed that measure P-450 activity (e.g., EROD) in a variety of species, including plants. With development of microplate protocols, the EROD bioassay has the dual advantages of being fast and cost-effective when large numbers of samples are analyzed (Whyte et al. 2000). With many biochemical indicators, P-450 activity assays are sensitive at lower contaminant concentrations and detect changes before the onset of gross adverse effects, as has been demonstrated in animal systems (Whyte et al. 2000). However, similar sensitivity has not been explicitly shown for plants as it has for animals.

Because P-450's, in plants, are involved in more diverse reactions, caution should be exercised in comparisons with animal systems. Several studies have demonstrated the utility P-450 assays as biomarkers of exposure, although responses (induction vs. inhibition) in these studies have been markedly variable. Researchers should consider that induction of P-450 enzyme activity represents the cumulative impact of all inducing chemicals taken up by the organism, regardless of whether or not they are detected analytically (Whyte et al. 2000).

4.5 Disadvantages

Because so many xenobiotics induce P-450 enzyme activity, there are limits to the diagnostic ability of assays such as EROD. Consequently, it is difficult to develop definitive cause-and-effect relationships for specific contaminants. Confounding of results can occur because many substrates may competitively bind to, or allosterically alter the structure of CYP1A (disguising the effects of other chemicals), particularly with mixtures. Inhibition of P-450 catalytic activity may affect the toxicity of chemicals through potentiation or antagonism, which may not be recognized when measuring EROD activity alone (Whyte et al. 2000). Whyte et al. (2000) points out that no exposure-response relationships have been developed in fish for use of EROD as a biomarker, and in plants no examples were found at the time of this writing. Another disadvantage of P-450 activity assays, such as EROD, is the availability of more specific and sensitive assays, including ones of gene expression (Whyte et al. 2000). Unlike enzymatic activity measurements, gene expression is not subject to the end product or substrate catalytic inhibition by contaminants, and allows measurement of P-450 induction where tissue samples are limited (Whyte et al. 2000). In addition to their role in the detoxification process, plant P-450s also participate in a myriad of biochemical pathways, including those devoted to the synthesis of plant products. Projections suggest that there are over 10,500 P-450's in higher plants (Nelson et al. 2004), portending exceptional diversity and difficulty in characterizing a given response.

5 Metabolic Enzymes: Phase II Metabolic Enzymes

5.1 Overview

Of the phase II metabolic enzymes, glutathione *S*-transferases (GSTs) show the most promise as biomarkers for xenobiotic stress. GSTs are a large group of related proteins present in eukaryotes that detoxify xenobiotics by conjugation with glutathione (Edwards and Dixon 2000). GSTs are constitutive enzymes in plants, and form part of the glutathione pathway. Glutathione is an antioxidant in cells, acts as a storage compound for sulfur, and may be an important signaling molecule

(Tausz et al. 2004). In addition to the constitutive forms, GST content and activity can be induced in response to various xenobiotics. Two studies, one with 59 species of lower plants and algae, and the other with 10 macrophyte species, found that GST activity was inducible by model xenobiotics in almost all tested species, despite the crude analytical methods used (Pflugmacher et al. 2000a; Pflugmacher and Steinberg 1997).

5.2 *Quantification*

GST activity can be measured spectrophotometrically at 340 nm, using the artificial substrate 1-chloro-2,4-dinitrobenzene (CDNB) to form a DNP-glutathione conjugate [DNP-SG] via nucleophilic displacement of Cl with the GSH-thiol (Mauch and Dudler 1993); other substrates can also be used (Pflugmacher and Schroder 1995; Pflugmacher et al. 2000a).

5.3 *Applications to Evaluate Xenobiotic Stress*

GST is induced by xenobiotics in a variety of species, and the range of chemicals that induce GST activity in plants is quite broad, encompassing the following: insecticides (methyl parathion), fungicides (folpet), herbicides (dichlofop-methyl, paraquat, pretilachlor, atrazine, fluorodifen), a variety of organic pollutants (PAHs, pentachlorophenol, pyrene, nitrobenzene derivates, TBTCl, hexachlorobenzene), natural toxins (cinnamic acid, cyanobacterial toxins) and metals (Al, Ni) (Anderson and Davis 2004; Deng and Hatzios 2002; Lei et al. 2006; Maya-Chavez et al. 2006; Mitrovic et al. 2004; Paskova et al. 2006; Pflugmacher 2004; Pflugmacher et al. 2000a, b, 2001; Roy and Hanninen 1994; Roy et al. 1995; Schrenk et al. 1998; Simonovicova-Olle et al. 2006; Teisseire and Vernet 2001). In *Lemna minor*, Teisseire et al. (2000) found that $CuSO_4$ stimulated GST activity up to 125% of controls, though activity decreased at higher concentrations, probably from cellular damage. Similarly, glutathione pools were quickly depleted by >50% in *Hydrilla verticillata* exposed to lead within 96 hr (Gupta et al. 1995). In contrast, *Lemna minor,* exposed to the rice herbicide propanil did not increase GST activity significantly at EC_{80} concentrations, during the first week of exposure (Mitsou et al. 2006). Similarly, exposure of *Typha latifolia* to methyl parathion did not consistently induce significant GST activity (Maya-Chavez et al. 2006). Differential responses were also found for *Phragmites australis* exposed to cyanotoxins, where microcystin induced a response but nodularin did not (Pflugmacher et al. 2001). In *L. minor,* it was proposed that propanil was metabolized primarily by a pathway other than conjugation with glutathione. Alternative pathways and differences in xenobiotic levels tested may account for the variable results, because studies with *L. minor* and *T. latifolia* did show increased GST, but not at significant levels (Maya-Chavez et al. 2006; Pflugmacher et al. 2001). For any given toxin GST activity can be regulated differently in different species. A study on four algal species exposed to pyrene showed large species-specific differences; the differences were related to differential compound-tolerance (Lei et al. 2006). For selective herbicides, it is known that selectivity among species derives largely from differences in the rate of pesticide conjugation with glutathione (Edwards and Dixon 2000).

GSTs are often measured for xenobiotics that tend to induce oxidative stress (Lei et al. 2006; Paskova et al. 2006; Pflugmacher 2004; Simonovicova-Olle et al. 2006), or in pesticide crop or weed metabolism studies (Deng and Hatzios 2002; Kreuz and Martinoia 1999; Sergiev et al. 2006; Zhao and Zhang 2006). Few such studies link induction of GST to growth parameters. For those that do, changes in GST activity appears to be a rapid biomarker; clear responses occur <24 hr after xenobiotic exposure, when levels applied exceed the EC_{50} for growth (Lei et al. 2006; Mitrovic et al. 2004). Lower, chronic exposure levels appear to induce GST and maintain the induction over time. *Myriophyllum quitense*, transplanted from a non-polluted area to rivers of varying nutrient- and heavy metal-pollution, showed a good statistical correlation between induction of GST, glutathione reductase and peroxidase and degree of pollution (Nimptsch et al. 2005). Similarly, *Nuphar lutea* showed increased GST activity when grown in PAH-polluted lakes (Schrenk et al. 1998). Therefore, GST activity, and perhaps other enzymes active in xenobiotic degradation such as peroxidases (see Sect. 9) (Jansen et al. 2004; Nimptsch et al. 2005) may yet prove to be sensitive indicators of xenobiotic pollution in the field.

5.4 Advantages

Inducible activity for GST, and other antioxidant/biotransformation enzymes, is both rapid and sensitive as a biomarker for a wide range of chemical pollutants. GST enzyme activity is less influenced by variations in plant-growth conditions, and is therefore more suitable for field evaluations. This is in contrast to, for example, pigment content and photosynthetic parameters (Sects. 10 and 11). The utility of GST activity is enhanced if the GSTs involved in ROS scavenging (likely to be affected by environmental stresses), and the inducible GSTs involved in detoxification of xenobiotics, can be distinguished. Recent studies illustrate that it may be possible to distinguish between GST types. Gene expression of five different GSTs, one glutathione peroxidase and one glutathione reductase was assessed in *Euphorbia esula* exposed to either the herbicide dichlorfop-methyl or to cold or drought stress. The expression patterns induced by the herbicide and drought stress were quite similar, but markedly different from the pattern observed for cold stressed plants (Anderson and Davis 2004). Zhao and Zhang (2006) compared the expression of GST, catalase (CAT) and superoxide dismutase (SOD) in transgenic rice exposed to either salt stress or to the herbicide paraquat. For salt and paraquat exposure, CAT and SOD activity increased, indicating oxidative stress. However, GST activity only increased significantly in the paraquat treated plants (Zhao and

Zhang 2006). This suggests that the measured GST was primarily involved in xenobiotic biotransformation, assuming the levels of oxidative stress in the two treatments were comparable. Further studies of different GSTs and their function may lead to specific biomarkers of xenobiotics exposure.

5.5 Disadvantages

The main disadvantage of GSTs is the dual roles they play in xenobiotic biotransformation and ROS scavenging. GST activity may respond to non-chemical causes of ROS formation such as drought and chilling stress, high ammonia concentrations and virus infections (Anderson and Davis 2004; Niehl et al. 2006; Nimptsch and Pflugmacher 2007; Tausz et al. 2004). However, such non-specific responses may be overcome by differentiation among inducible and constitutive forms of GSTs. Although prohibitive if all species were included, differentiation among various GSTs is realistic, if executed on selected cosmopolitan species. *Lemna spp.* and *Myriophyllum spp.* could be selected for monitoring freshwater and sediment quality, respectively, and macroalgae such as *Ulva spp.* may be a good candidate for field monitoring of saltwater quality. Using a few, thoroughly investigated, species would also overcome the problem of variation in regulation of GST between species (Lei et al. 2006). Of course, measuring enzyme activity is not as easy as measuring fluorescence or biomass growth, for example. Notwithstanding, there are several standardized methods to measure GST activity that only requires a spectrophotometer (Pflugmacher et al. 2000a); hence, measurements can be performed at reasonable cost by most laboratories.

6 Phytochelatins

6.1 Overview

Phytochelatins (PCs) are a class of heavy-metal binding peptides consisting of repetitive γ-glutamylcysteine units with a carboxyl-terminal glycine, ranging from 5 to 17 amino acids in length. PCs are synthesized by all plants and exist in sizes varying from 2 to 11 γ-gltutamylcysteine units (Grill et al. 1987). In plants, PCs are involved in detoxification and homeostasis of heavy metals, and are functionally analogous to metallothioneins in animals and some fungi (Grill et al. 1987). To synthesize PCs, the enzyme γ-glutamylcysteine dipeptidyl transpeptidase (phytochelatin synthase) catalyzes the transfer of the γ-glutamylcysteine dipeptide moiety of glutathione to an acceptor glutathione molecule or a growing chain of [Glu(-Cys)],-Gly oligomers (Grill et al. 1989). γ-glutamylcysteine is also a precursor for glutathione (Grill et al. 1987). Phytochelatin synthase (PS) is constitutively present in plant cells. Although it's formation is not noticeably induced by heavy metal ions, heavy metals such as Cd^{2+}, Ag^{+}, Bi^{3+}, Pb^{2+}, Zn^{2+}, Cu^{2+}, Hg^{2+}, and Au^{+} increase the activity of phytochelatin synthase in a self-regulating loop (Grill et al. 1989). Comparatively, no enzyme activation has been detected with other metals such as $A1^{3+}$, Ca^{2+}, Fe^{2+}, Mg^{2+}, Mn^{2+}, Na^{+} and K^{+} (Grill et al. 1989). The synthesis of the individual PCs, or iso-phytochelatin species, is governed by the availability of the specific GSH or C-terminal-modified GSH molecules, within a given plant (Zenk 1996). It is essential that PCs chelate and inactivate toxic metal ions entering the cytosol, before they can poison the enzymes of life-supporting metabolic routes (Zenk 1996).

6.2 *Quantification*

To analyze for this enzyme, plant tissues are extracted and processed as outlined by Grill et al. (1987), separated by HPLC, derivitized at sulfhydryl groups and detected at 410 nm. A more recent and refined method is provided in Srivastava et al. (2006). Assay methods for the extraction, purification and activity of phytochelatin synthase are outlined in Grill et al. (1989).

6.3 *Applications to Evaluate Xenobiotic Stress*

Several studies have shown that PCs are induced after exposure to metals. This induction represents a unique biomarker of metal exposure and, potentially, of metal effect. Grill et al. (1987) demonstrated that PCs are selectively induced in *Rauvolfia serpentina* (indian snakeroot) cell suspension cultures; the ions most active in inducing heavy-metal-binding molecules were Cd^{2+}, Zn^{2+}, Pb^{2+}, Ag^{+} and Sb^{3+}. Induction was also observed after exposure to Ni^{2+}, Au^{+}, Bi^{3+}, Sn^{2+}, and the anions AsO_4^{3-} and SeO_3^{2-}, which were not known to induce metallothionein synthesis in humans and animals (Grill et al. 1987). Furthermore, a range of metal salts including $A1^{3+}$, Ca^{2+}, Co^{2+}, Cr^{2+}, Cs^{+}, K^{+}, Mg^{2+}, Mn^{2+}, MoO_4^{2-}, Na^{+} or Va^{2+} were found not to cause induction in the same test system (Grill et al. 1987).

Cysteine, GSH, PC2 (a dimer) and PC3 (a trimer) were all found in the submerged macrophyte *Hydrilla verticillata* after exposure to copper; both PC2 and PC3 were induced by 5.6- and 9.4-fold, compared to controls, respectively (Srivastava et al. 2006). Higher exposures resulted in further PC2 and PC3 increases by another three- and twofold, respectively (Srivastava et al. 2006). Phytochelatins were also induced in *H. verticillata* exposed to lead (2.5 μM), where PC2 and PC3 concentrations increased to greater than twice the control levels (Gupta et al. 1995).

Pawlik-Skowrońska (2001) evaluated the production of PCs in two freshwater filamentous green algae (*Stigeoclonium spp.*) that were exposed to mining water containing a mixture of heavy metals (17 μM). Results showed that both species accumulated similar levels of PCs at an alkaline pH (8.2): 500–600 nmol SH g^{-1} dry weight. However, acidified conditions (pH 6.8) resulted in a >2-fold increase of total PC levels, and the appearance of a longer chain peptide (PC4) in response to a fourfold increase in the concentration of labile forms of zinc (Pawlik-Skowrońska 2001). Individual metals (Cd, Pb and Zn) induced higher amounts of PCs (PC2–PC4) than the metal mixture (Pawlik-Skowrońska 2001).

Yin et al. (2002) found that PC2 and PC3 concentrations increased to 658 and 118 nmol/g fresh wt, respectively, from 0 in controls of *Lemna aequioctialis* exposed to 20 μM of cadmium. By comparison, the colony count for *L. aequioctialis* decreased by only 10.5%, suggesting greater sensitivity for PC content analyses. Confirming the specificity of PCs for metals, when buthionine sulfoxine (BSO; a potent inhibitor of phytochelatin synthase) was added in combination with cadmium, a dramatic inhibition of growth occurred (Yin et al. 2002).

6.4 Advantages

Phytochelatins are highly specific for heavy metals and can facilitate identification of heavy metal exposure when mixtures of contaminants exist. Inductions of PC concentrations show good agreement with increasing heavy metal exposure, which allows for concentration-dependent evaluations. PC levels have demonstrated comparable sensitivity to growth measures, while providing more specific diagnostic resolution. Test methods that use PCs are fairly refined and have been employed for two decades; favorable past experience with PCs has conferred confidence in methods and markers.

6.5 Disadvantages

Although PCs are excellent biomarkers for heavy metal exposure, their activity is generic among this class of contaminants. Therefore, they are inadequate to diagnose which heavy metal is present, or is causing effects. Measuring PC levels requires considerable sample preparation and rather sophisticated analytical equipment.

7 Flavonoids

7.1 Overview

Flavonoids (phenylpropanoids) are a class of phenolic natural products comprising more than 4,500 compounds. They are found in most plant tissues as monomers, dimers and oligomers, and serve diverse functions ranging from pollinator attractants to UV-B screens (Croteau et al. 2000; Iwashina 2000). Flavonoids consist of various groups of plant metabolites including chalcones, aurones, flavonones, isoflavonoids, flavones, flavonols, leucoanthocyanidins, catechins and anthocyanins (Croteau et al. 2000). The first committed step of the flavonoid biosynthetic pathway is catalyzed by chalcone synthase, and several stress-induced phenylpropanoids are derived from the resultant C15 flavonoid skeleton (Dixon and Paiva 1995). Stimulation of phenylpropanoid production has been documented for a variety of environmental stresses, including high light/UV, pathogen attack, wounding, low temperatures as well as deficiencies in nitrogen, phosphate and iron (Dixon and Paiva 1995). Currently, the only major class of environmental contaminants shown to elicit effects on phenylpropanoid production is the heavy metals (Babu et al. 2003; Kidd et al. 2001). However, there is little information on how these stressors signal the accumulation of specific flavonoids (Babu et al. 2003); examples of flavonoid induction from xenobiotic exposure are limited.

7.2 *Quantification*

7.2.1 Chalcone Synthase

Total chalcone synthase protein content can be analyzed using sodium dodecyl sulfate polyacrylamide gel electrophoresis (SDS-PAGE) separation, and immunoassay (Babu et al. 2003). The antibody-chalcone synthase complexes are detected with peroxidase-conjugated rabbit anti-goat IgY; the chemiluminescence produced by the secondary antibody is detected using radiographic film (Babu et al. 2003).

7.2.2 Flavonoids

Current methods for extraction of flavonoids involve either heat drying (Miean and Mohamed 2001) or freeze drying (Michael et al. 1992) plant tissue samples, followed by solvent and acid extraction. The resultant flavonoid aglycons are quantified using reversed-phase HPLC and UV detection at 365 nm (Michael et al. 1992; Miean and Mohamed 2001).

7.3 *Applications to Evaluate Xenobiotic Stress*

Babu et al. (2003) found that chalcone synthase was up regulated by exposure to a combination of copper and photosynthetically active radiation (PAR) (400–700 nm), and simulated solar radiation (SSR) (containing UV-B: 290–320 nm) alone. In both cases, the sensitivity of chalcone synthase induction was proportional to that of plant growth rate. Compared to plants grown under standard PAR, plants exposed to SSR experienced nearly a 200% induction of chalcone synthase (Babu et al. 2003).

Similarly, plants exposed to 8 μM $CuSO_4$ under PAR conditions also produced nearly a 200% induction of chalcone synthase (Babu et al. 2003). Exposure to copper in combination with PAR, and SSR-alone also caused accumulation of virtually the same population of phenylpropanoid compounds. Flavonoids are known to act as scavengers of ROS, and both copper and UV generate ROS via independent mechanisms (Babu et al. 2003). Moreover, some hydroxyl group-containing flavonoids can chelate heavy metals. Thus flavonoids show promise as biomarkers of heavy metal exposure.

7.4 Advantages

Although specific induction patterns, or signatures, have not been developed for groups of contaminants, current research suggests some commonality in signal transduction and response pathways for ROS-forming stresses (see Sect. 9). To date, only heavy metals have been shown to induce flavonoid biosynthesis. Other ROS-facilitating contaminants may be appropriately evaluated via flavonoid analysis, however this is speculative. Because copper induces chalcone synthase to a similar degree and in the same manner as UV exposure (ROS formation), induction of this enzyme appears to be a useful biomarker for heavy metal exposure. Thus flavonoid analyses may contribute to further understanding and characterization of toxic response mechanisms.

7.5 Disadvantages

To date, little has been done on the specificity of flavonoid induction to particular groups of contaminants. Therefore, it is not conclusive whether flavonoid induction may serve as a biomarker of general stress, be specific to a given stressor type, or both. The work required to extract and analyze flavonoid induction are laborious, and enzyme-specific evaluations of chalcone synthase require specific antibodies. Because flavonoids serve such diverse functions in plants, their response to xenobiotic exposure may be difficult to interpret.

8 Stress Proteins (Heat Shock Proteins)

8.1 Overview

Lewis et al. (2001) states that all organisms respond to stress at the cellular level through rapid synthesis of a few "stress proteins," with associated simultaneous inhibition of normal protein synthesis. Initially described by Ritossa (1962) as heat shock proteins (HSPs) produced in *Drosophila melanogaster* cells following exposure to high temperature, a diverse range of environmental stresses have subsequently been identified, precipitating the more generalized "cellular stress response (CSR)" designation (Lewis et al. 2001). Exposure to trace metals, organic pollutants, UV radiation, changes in temperature or osmolarity, hypoxia and anoxia are all environmental stresses which can induce these proteins (Lewis et al. 1999). However, stress protein research has been confined almost exclusively to animals, despite the widespread use of plants as pollution monitors (Lewis et al. 2001).

Although some stress proteins are only found in cells that are responding to stressors, most are present at lower concentrations under normal conditions (constitutively expressed) and play essential roles in cellular protein homeostasis by acting as molecular chaperones (Lewis et al. 2001). When stress events occur, levels of constitutive stress proteins rise and inducible forms begin to be synthesized (Lewis et al. 2001). It is thought that under stress conditions, stress proteins take on additional but related functions to molecular chaperoning, helping repair denatured proteins and protecting others from damage (Lewis et al. 2001; Waters et al. 1996). Such protection allows cells to recover and survive the stress. As a primary protective response of cells, CSR is potentially useful in environmental monitoring (Lewis et al. 2001; Sanders 1990). Stress-proteins can be divided into families according to molecular weight; four major stress-protein or HSP families of 90, 70, 60 and 16–24 kDa are prominent, and are, respectively referred to as hsp90, hsp70, hsp60 and low molecular weight (LMW) stress-proteins (Bierkens 2000). A large physiological cost is thought to be associated with stress protein production (Krebs and Loeschcke 1994), and HSP accumulation may represent 10–15% of total soluble protein in tissue (Lewis et al. 2001).

8.2 *Quantification*

The approaches most frequently used to quantify stress proteins include metabolic labeling, followed by autoradiography and immunoassays with protein-specific antibodies or cDNA probes (Lewis et al. 1999). Western blotting, slot blotting, radioimmuno-assays (RIAs) and enzyme linked immunosorbant assay (ELISAs) are all methodologies employed to detect and quantify stress proteins (Lewis et al. 1999). An Hsp70 ELISA has been optimized for the green algae *Pseudokirchneriella subcapitata* (formerly *Selenastrum capricornutum*), using a monoclonal antibody with broad cross-reactivity for Hsp70 among species, and sensitivity gauged at >2 ng (Bierkens et al. 1998). Bierkens et al. (1998) suggest that measuring Hsp70 induction after 3 d rather than 6 d of exposure, may increase test sensitivity. Ireland et al. (2004) have also developed an indirect competitive ELISA (IC-ELISA) with a working range of 0.025–10 mg/L; this assay employs a monoclonal antibody raised against purified Hsp70 of *Phaseolus aureus* (mung bean) for use in marine macroalgae and fresh water plant species.

8.3 Applications to Evaluate Xenobiotic Stress

The use of HSPs as biomarkers in plants has been assessed in a few studies, with mixed results. In *Enteromorpha intestinalis* (a seaweed) exposed to copper and an antifouling herbicide (Irgarol 1051; a triazine herbicide), growth was the most sensitive measure of exposure in both cases, whereas, the Hsp70 response was comparatively weak (Lewis et al. 2001). Lewis et al. (2001) suggest that the relative sensitivities of each endpoint are related to the contaminant's mode of action; Hsp70 is only induced by stressors that are strongly proteotoxic. Copper is known to bind to proteins as a primary mode of action by interacting with sulphydryl groups of intracellular proteins, and disrupting their conformation (Lewis et al. 2001). Under nutrient-replete conditions, copper did induce an increase in Hsp70 levels. Paraquat is the only herbicide reported to induce stress proteins (Sanders 1993), and is potentially highly proteotoxic, because it facilitates the formation of ROS during photosynthesis (BPCP 2002). Triazine herbicides, such as Irgarol 1051, are thought to produce free radicals as a secondary consequence of blocking photosynthesis (Ahrens 1994), although, to a lesser extent than does paraquat.

In tests with the green algae *R. subcapitata,* stress-proteins demonstrated greater sensitivity than did growth inhibition endpoints, when exposed to different classes of environmental pollutants including heavy-metals (Zn^{2+}, Se^{2+}), carbamates (carbaryl) and organochlorine insecticides (lindane), chlorinated hydrocarbon insecticides (pentachlorophenol; PCP), and surfactants (sodium dodecyl sulfate; SDS) (Bierkens et al. 1998). With the exception of PCP, all stressors evaluated caused a concentration-dependent induction of Hsp70 synthesis. The heavy metals Zn^{2+} and Se^{2+} induced Hsp70 synthesis at concentrations one to two orders of magnitude below the concentrations at which effects on growth are observed (Bierkens et al. 1998). Heavy metals are among the most potent inducers of Hsp70 at concentrations equal to or approaching the limits set by regulatory authorities (Bierkens et al. 1998). Furthermore, SDS, carbaryl and lindane induced Hsp70 at concentrations that caused only minimal growth inhibition in *R. subcupituta.*

Ireland et al. (2004) evaluated Hsp70 production in response to cadmium exposure in three species of marine macroalgae (*Fucus serratus*, *Chondrus crispus*, *Ulva lactuca*), and one species of duckweed (*Lemna minor*). Exposure of *F. serratus* to cadmium resulted in a concentration-dependent increase in Hsp70 levels up to tenfold that of control values, beyond which Hsp70 concentrations decreased (Ireland et al. 2004). Similarly, *L. minor* showed a concentration-dependent increase in Hsp70 levels of more than fivefold, but subsequently decreased to levels equal to control values. The reductions in Hsp70 concentrations above certain stress thresholds may be attributed to cellular damage.

Lewis et al. (2001) suggest that an antibody that detects only the inducible form of Hsp70, instead of total Hsp70 (constitutive and inducible forms) may confer a more specific and sensitive response. Moreover, Hsp60 may be more sensitive to copper toxicity than Hsp70, because it is found primarily in mitochondria and chloroplasts, which are the sites of redox reactions and involve many metallo-enzymes that copper may bind to and disrupt. LMW stress proteins are known to be particularly abundant and responsive to stress in higher plants (Waters et al. 1996), and may provide a more sensitive response to certain stressors (Lewis et al. 2001). This may also hold true for algae.

8.4 Advantages

Stress-proteins offer an advantage over analytical methods in that they measure the effective fraction of pollution that affects an organism; they also effectively integrate multiple exposure routes for a given interval and number of pollutants (Bierkens 2000). Furthermore, because stress-proteins are integrators of protein damage, they may provide added value to biomonitoring, by increasing the sensitivity of biological testing and complementing tests that measure more specific modes of toxic action (Bierkens 2000; Bierkens et al. 1998; Sanders et al. 1994). Hsp70 is induced by a wide variety of chemicals, and evidence exists that measuring Hsp70 induction adds to detection sensitivity by a factor of 5–1,000, depending on the class of chemicals involved (Bierkens et al. 1998). Because heat-shock response occurs under conditions which constrain cell metabolism, Bierkens (2000), and Meyer et al. (1995) concluded that it does not represent a sensitive marker for subtle pharmacological or toxicological effects, but contributes to a better estimation of cell damage at denoted levels of protein synthesis inhibition and/or proteotoxicity. Similarly, de Pomerai (1996) believes that, for mixtures, stress-proteins constitute an integrated response of the total damage caused to proteins within the organism.

8.5 Disadvantages

Although stress proteins may have increased sensitivity over morphological endpoints (Bierkens et al. 1998), different agents induce different families of stress-proteins with widely differing efficiencies. Therefore, screening for only one class of stress-proteins may not provide a sufficiently sensitive bioindicator for a range of pollutants (Bierkens 2000; de Pomerai 1996). Furthermore, stress proteins may be a less sensitive biomarker than those which quantify and characterize effects at the primary target, because proteotoxicity, resulting in Hsp activation, may be a secondary consequence of toxicological damage (de Pomerai 1996). Thus, stress proteins appear to be less sensitive for contaminants that are not primarily proteotoxic (Lewis et al. 2001). The stress protein response can also be influenced positively or negatively by a variety of xenobiotics, thereby complicating the interpretation of data from field studies in which xenobiotic mixtures are present (Bierkens 2000). In addition, HSPs are induced by changes in temperature (Ireland et al. 2004), and inhibited under nutrient- limiting conditions (Lewis et al. 2001), which may further confound the response and render environmental control difficult. High variability in response is also an issue.

9 Reactive Oxygen Species and Scavenging Enzymes

9.1 *Overview*

The generation of, and protection against ROS is an intrinsic characteristic of any living cell (Bolwell and Wojtaszek 1997). ROS are part of the alarm-signaling processes in plants and serve to modify metabolism and gene expression that allow plants to respond to adverse environmental conditions, invading organisms and UV irradiation (Foyer et al. 1994). In plants, the production and destruction of ROS is intimately involved with the regulation of photosynthetic electron flow, where stress-mediated imbalances may induce ROS, and lead to photoinhibition and photooxidation of the organelles in the chloroplast (Foyer et al. 1994). The photosynthetic electron transport system is the major source of ROS in plant tissues. ROS have the potential to generate singlet oxygen $^1O_2$ and superoxide $O_2^{\bullet -}$, which in turn, can be successively reduced to hydrogen peroxide (H_2O_2) and hydroxyl radical (•OH) (Foyer et al. 1994). ROS formation in plants stimulates defense mechanisms in which antioxidant systems are induced to detoxify different ROSs, and assist in acclimating plants to oxidative stress (Babu et al. 2003; Dixon et al. 1998; Foyer et al. 1994). For example, superoxide is detoxified by SOD, hydrogen peroxide is detoxified by ascorbate peroxidase (APX), and hydroxyl radicals are scavenged by glutathione through GST (Babu et al. 2003; Bowler et al. 1992; Dixon et al. 1998; Foyer et al. 1994). Thus, up-regulation of a particular antioxidant system in stressed plants may reveal the nature of ROS formed (Babu et al. 2005). Both natural and man-made stress can provoke increased production of toxic oxygen derivatives. Oxidative damage arises in high light, principally in conjunction with additional stress factors such as chilling temperatures or pollution (Foyer et al. 1994). Under non-stressful conditions, the antioxidative defense system provides adequate protection against the background production of both reactive oxygen and free radicals produced under normal light conditions. During a stress event, which induces additional ROS formation, the capacity of the antioxidative defense system is often increased. If this increase is insufficient, ROS production will exceed scavenging capacity and lead to metabolic disruption, loss of function, and tissue destruction (Foyer et al. 1994). Evaluations of ROS and antioxidant enzymes are most appropriate for compounds that target the photosynthetic chain by disrupting the flow of electrons, particularly metals and certain organics. ROS production and antioxidant enzymes can potentially be used to identify the source (type/nature) of stress.

9.2 *Quantification*

9.2.1 Reactive Oxygen Species

Production of ROS in *Lemna gibba* has been measured using H_2DCFDA (2′,7′-dichlorodihydrofluorescein diacetate), a non-fluorescent compound that is readily taken up by cells (Babu et al. 2003; Behl et al. 1994). Once inside, the acetate is cleaved by endogenous esterases and the acetate-free reduced form (H_2DCF) becomes trapped inside the cells, and can be oxidized by ROS, particularly the hydroperoxides, to form a highly fluorescent fragment of H_2DCFDA, DCF (Babu et al. 2003). DCF fluorescence is measured using an excitation wavelength of 485 nm and an emission wavelength of 530 nm, and measurements can be made both on intact plants and homogenates (Babu et al. 2003).

9.2.2 Plant Extract Preparation

Crude plant extracts are prepared according to the methods of Jansen et al. (1996). Antioxidant activities in the supernatants can then be assayed for all treatments on an equal protein basis, as confirmed using a Bradford reagent (Babu et al. 2005).

9.2.3 Superoxide Dismutase

As outlined by Babu et al. (2003) and modified from Beauchamp and Fridovich (1971), SOD activity can be measured using non-denaturing polyacrylamide gel electrophoresis. Activity is assessed using nitroblue tetrazolium (electron acceptor for colorimetric and spectrophotometric activity assays of oxireductases), which appears as a negative stain under cool white fluorescent light. The SOD isozymes are identified by inhibitor studies, as described by Bowler et al. (1992), and the density of each SOD isozyme is quantified by densitometry.

9.2.4 Ascorbate Peroxidase

The oxidation of ascorbic acid by APX is measured spectrophotometrically as the decrease in absorbance of the substrate at 290 nm (Nakano and Asada 1981). The assay medium contains Na-ascorbate and a known amount of protein; the reaction is started by the addition of hydrogen peroxide. Correction can be made for the low, non-enzymatic oxidation of ascorbate by hydrogen peroxide (Teisseire and Guy 2000).

9.2.5 Glutathione Reductase

Glutathione reductase (GR) is a ubiquitous NADPH-dependent enzyme, which catalyzes the reduction of oxidized glutathione (GSSG) (Smith et al. 1988). Glutathione reductase activity is measured spectrophotometrically (Smith et al. 1988). The assay medium contains 5,5′-dithiobis-2-nitrobenzoic acid (DTNB) and oxidized GSSG; GR activity is measured as a rise in absorbance at 412 nm from the reduction of DTNB by GSH to TNB (2-nitro-5-thiobenzoic acid) (Smith et al. 1988).

9.3 Applications to Evaluate Xenobiotic Stress

The effects of contaminants on ROS production and scavenging enzymes have been investigated in several studies, particularly metals and organics that target the photosynthetic chain with consistent results. Teisseire et al. (2000) found that copper caused significant modification of GR activity in *Lemna minor*. Concentration-dependent increases of pyrogallol peroxidase, guaiacol peroxidase and catalase activities were observed at all treatment concentrations; catalase and guaiacol peroxidase, in particular, were strongly induced at levels as high as 347% and 374% of the basal level, respectively (Teisseire and Guy 2000). Similar to GR and GST, APX activity increased to a maximum of 132% compared to controls, though beyond this concentration APX activity decreased (Teisseire and Guy 2000). Teisseire et al. (2000) concluded that, because of their sensitivity, certain antioxidant enzymes may be useful as water quality biomarkers. Unfortunately, growth parameters were not reported as a means of comparison.

Babu et al. (2005) evaluated the effects of copper and 1,2-dihydroxyanthraquinone (1,2-dhATQ; a PAH), alone and in combination in *L. gibba*. Singularly, copper and 1,2-dhATQ caused 7% and 35% inhibition of growth at 4 mM and 3 mM, respectively; however, the combination produced a synergistic inhibition of growth rate (70%) (Babu et al. 2005). Comparatively, treatment with copper alone and in combination with 1,2-dhATQ, induced Mn SOD and Cu–Zn SOD isoforms by ~50–200%, indicating that certain isoforms of SOD can provide a two to threefold more sensitive effect measure than growth (Babu et al. 2005). Exposure to copper and 1,2-dhATQ, in combination, caused an increase in ROS production by ~75% and a 2.5-fold induction of GR activity, compared to controls; singular treatments had no effect on either parameter (Babu et al. 2005). In contrast, APX activity was induced upon exposure to copper, unchanged upon exposure to 1,2-dhATQ and decreased upon exposure to the binary combination (Babu et al. 2005). The exogenous addition of DMTU (dimethyl thiourea, a potent ROS scavenger) was found to alleviate the formation of ROS in plants subjected to the mixture of copper and 1,2-dhATQ, suggesting that H_2O_2 and •OH were the causative ROS agents, because DMTU is a specific scavenger of both radicals (Babu et al. 2005).

Babu et al. (2003) also evaluated the simultaneous effects of copper and UV radiation, in the form of SSR (a light source containing PAR and UV radiation), in *L. gibba*. Both copper and SSR caused ROS formation, although the ROS levels were higher when copper was combined with SSR than when applied with PAR. In comparison to PAR grown plants, SSR treated plants exhibited elevated levels of SOD and GR, and these enzyme levels were further elevated when copper was added at concentrations that generated ROS (Babu et al. 2003).

Hence, evaluating ROS production directly, as well as the induction of scavenging enzymes, provides a highly specific biomarker for ROS-mediated toxicity, and shows greater sensitivity than growth for certain metrics (SOD).

9.4 Advantages

Differential antioxidant responses to specific chemical treatments suggest that the formation of different ROS species can vary greatly between contaminants. Induction or suppression of certain antioxidant enzymes can disclose which types of ROS are responsible for a given effect. Evaluation of ROS production and antioxidant enzymes may also portend a specific mechanism of action. Furthermore, ROS production and antioxidant levels correlate logically and strongly with effects on growth (Babu et al. 2003, 2005). Several antioxidant assays provide more specific and sensitive indicators of stress when compared to morphological and pigment endpoints (Sect. 10) (Babu et al. 2005). With the exception of SOD, measuring ROS production and antioxidant levels does not require exhaustive sample preparation and specialized equipment, because most assays can be measured spectrophotometrically.

9.5 Disadvantages

Although evaluation of antioxidant enzymes provides insight into the causative ROS that contributes most to a particular response, the responses are complex and currently cannot be attributed to a single ROS (Babu et al. 2005). Evaluating the levels of SOD transcripts requires extensive sample preparation and specialized equipment, because quantification requires densitometry analysis and isozyme identification requires conducting inhibitor studies. The use and application of ROS production and antioxidant induction/suppression is largely untested under field conditions. This introduces uncertainties and confounding in responses because of the complexity of environmental matrices and mixtures.

10 Plant Photosynthetic Pigments

10.1 Overview

Chlorophylls and carotinoids are the primary light capturing pigments in higher plants, and are located in thylakoid membranes of the chloroplast. Pigments function to absorb light energy for photosynthesis, and protect the photosynthetic apparatus from excess light. Excess light can create a surplus of excited electrons, which exceeds the capacity of the photosynthetic electron transport chain, leading to the formation of ROS (Buchanan et al. 2001). Light intensity, spectral composition of light, nutrient status and temperature are all factors that affect the content and composition of photosynthetic pigments, which vary naturally among species and with leaf position (Murchie and Horton 1997; Nielsen 1993). The main principle behind acclimation of pigment content to light, in plants, is optimization of light absorption under given conditions, and adjustment to the carbon fixation capacity of the plant. Hence, under conditions of low irradiance pigment content is increased to maximize light absorption; at high irradiance pigment content is reduced to avoid ROS damage (Murchie and Horton 1998). At high nutrient availability (particularly nitrogen), pigment content increases to enhance carbon fixation (Cedergreen and Madsen 2003; Eloranta et al. 1988). When temperature affects carbon fixation rates, the pigment content adjusts accordingly (Barko 1983; Eloranta et al. 1988). Particularly in the aquatic environment, light quality and quantity varies substantially with water color and depth (Wetzel 1983). Hence, even in the same stand of plants, leaf pigment content and composition will vary with water depth.

Xenobiotics may affect pigment synthesis directly, e.g., herbicides that target chlorophyll and carotinoid biosynthesis by inhibiting protoporphyrinogen oxidase and *p*-hydroxyphenyl pyruvate dioxygenase, respectively (Copping and Hewitt 1998). Other xenobiotic stressors can affect pigment content indirectly; for example, highly lipophilic chemicals acting to induce necrosis (Donkin 1994), or herbicides that interrupt membrane synthesis (Copping and Hewitt 1998). The herbicides diquat and paraquat, which disrupt the photosynthetic electron transport chain, may directly increase ROS formation and indirectly decrease the carbon sink capacity. Ultimately, disruption of electron flow leads to an excess of electrons at the light harvesting complex of the photosynthetic apparatus (Copping and Hewitt 1998; Marshall et al. 2000). Heavy metals are also known to induce ROS formation (Babu et al. 2005; Lagriffoul et al. 1998; MacFarlane 2003; Mocquot et al. 1996). Therefore, many chemicals, with different modes of action, may affect pigment composition.

10.2 Quantification

Photosynthetic pigments are typically measured spectrophotometrically after extraction with organic solvents such as acetone, ethanol, methanol or diethyl ether (Greenberg et al. 1992; Lichtenthaler 1987; Porra 2002). Absorbance of whole plant extracts is calculated from the ratio of extract to reference blank, using equations from Porra et al. (1989), Greenberg et al. (1992), Rowan (1989) and Ritchie (2006). More elaborate analyses, particularly on micro algae communities, where the community structure can be quantified through pigment content, are conducted on methanol extracts using HPLC (Marinho and Rodrigues 2003; Wilhelm et al. 1995).

10.3 Applications to Evaluate Xenobiotic Stress

A range of studies, primarily on heavy metals (Appenroth et al. 2001; Lagriffoul et al. 1998; Powell et al. 1996; Qi et al. 2006), PAHs (from creosote) (Marwood et al. 2001) and herbicides (Cedergreen et al. 2007; Frankart et al. 2003; Kushwaha and Bhowmik 1999; Turgut and Fomin 2002; Turgut et al. 2003), have shown chlorophyll and carotinoid content to be as or more sensitive stress indicators than biomass or relative growth rate. For example, measuring chlorophyll-a, -b, and carotinoid content in cucumber cotyledons treated with isoxaflutole, demonstrated nearly double the sensitivity compared to fresh weight (Kushwaha and Bhowmik 1999).

However, there are exceptions to this pattern of comparative sensitivity. Investigations of the effects of monochloroacetic acid and dichloroacetic acid on the aquatic macrophytes *Myriophyllum spicatum, M. sibiricum* and *Lemna gibba* showed little or no response for chlorophyll a, b or carotenoids at concentrations that gave good concentration-response relationships for morphological and growth-related endpoints (Hanson et al. 2002, 2003). Microcosm evaluations with several pharmaceuticals tested alone or in combination, have also shown this trend for *L. gibba, M. spicatum* and *M. sibiricum* (Brain et al. 2004a, 2005a, b). Laboratory exposures of *L. gibba* to several pharmaceuticals also showed that pigments were comparatively less sensitive than growth measures (Brain et al. 2004b). However, among pigments, chlorophyll b was generally more sensitive than chlorophyll a, whereas, carotenoids were the least sensitive endpoint (Brain et al. 2004b). The triazine herbicide terbuthylazine showed no effect on pigment content in *Lemna minor* at concentrations that completely arrested growth, even though ROS formation was expected (Cedergreen et al. 2007). There are two causes for this: some chemicals may arrest growth before pigment content is affected because of their mode of action; alternatively, the chemical's effect on pigments may be influenced by environmental growth conditions. In reality, it is probably a combination of both. Herbicides such as terbuthylazine, glyphosate and metsulfuron-methyl will never show the same level of immediate bleaching, even under full sunlight, as will the herbicide diquat, which can turn leaves yellow within one d. On the other hand, all three herbicides do show more bleaching under field conditions than they do when grown in growth chambers under moderate light intensities (80–200 μmol m^{-1} s^{-1} (PAR)) (Cedergreen et al. 2007, personal observations).

Pigment content and/or composition is generally used as a biomarker in the field for studies on micro-algae, where pigment content equates to living algae biomass, and pigment composition is indicative of species (Wilhelm et al. 1995). Field studies on pigment content as a biomarker in higher plants are rare. MacFarlane (2003) attempted to relate chlorophyll-a, -b and carotinoid content to sediment pollution and leaf accumulation of the heavy metals Cu, Pb and Zn, in mangrove (*Avicennia marina*). However, only a correlation between leaf Zn content, and chlorophyll a/b ratio was found, though this correlation was not maintained over time (MacFarlane 2003). Because only biochemical parameters were measured, it is difficult to relate these measures to ecologically relevant parameters such as growth, reproduction and survival (MacFarlane 2003). Regardless, the study reveals that when pigments show a correlation with pollutant exposure, the stability of this effect measure can be obscured or diminished by environmental factors. Therefore, pigments are not reliable measures of chemical pollution, in the field, unless appropriate controls are available for comparison.

10.4 Advantages

Pigment content can be an easy-to-measure and robust biomarker, amenable to both laboratory and field-based investigations. Inhibition of pigment content may portend modes of action, if the contaminant disrupts photosynthesis or pigment biosynthesis. Pigment content can be a more sensitive effect indicator than growth, particularly where the pigment biosynthetic pathway or photosynthetic apparatus is targeted directly (bleaching herbicides and PSII inhibitors). Greenberg et al. (1992) proposed that pigment content be conjoined with morphological endpoints, because pigments indicate damage to processes other than cell division. Furthermore, visual observation of plant coloration may preclude the necessity of measuring pigment content. Chlorophyll measures are frequently employed with algae, as an alternative to cell counts, to assess stress, because it is equally sensitive but easier to assay.

10.5 Disadvantages

Pigment content biomarkers have variable sensitivity compared to gross morphological endpoints, depending on the contaminant mode of action. Because they are not as contaminant-specific, pigment content endpoints are not as widely employed, as biomarkers of xenobiotic stress, as are morphological endpoints. The plasticity of pigment content and composition, in response to environmental conditions (particularly light for submerged plants), renders it less useful as a biomarker under field conditions.

11 Photosynthesis: Chlorophyll Fluorescence and Gas Exchange

11.1 Overview

Measurements of chlorophyll fluorescence are now widely used for studying plant photosynthesis and stress response (Force et al. 2003; Frankart et al. 2003; Marwood et al. 2001, 2003; Oxborough 2004; Roger and Weiss 2001; Strasser et al. 2000). Chlorophyll fluorescence will change in response to chemical contaminant exposure and abiotic stress (drought, suboptimal temperatures and nutrient deficiency) (Chesworth et al. 2004; Christensen et al. 2003; Deltoro et al. 1999; Kellomäki and Wang 1999; Mahan et al. 1995; Sayed 2003). Chlorophyll fluorescence generates a range of parameters that may provide information on the efficiency of the photosystem II (PSII) electron transport chain. The efficiency of PSII correlates well with CO_2 fixation, although this correlation often breaks down under different forms of stress (Maxwell and Johnson 2000). Direct measurements of net CO_2 fixation are necessary to determine photosynthetic rates, and thereby, the potential for net plant growth. Gas exchange measurements are more laborious than those for fluorescence and are seldom used to monitor xenobiotic or environmental stress.

11.2 Quantification

There are several different approaches to measure chlorophyll fluorescence. The simplest, quickest and most widely used method, particularly in the field, is the measurement of fast chlorophyll-a fluorescence induction. This method measures the variance in fluorescence of dark adapted plants/leaves, exposed to a supersaturating light flash that produces the so-called Kautsky curve (Force et al. 2003; Kautsky et al. 1960; Strasser et al. 2000).

Another often used method, particularly for laboratory studies, is pulse amplitude modulation (PAM) (Genty et al. 1989; Maxwell and Johnson 2000). The maximum quantum yield parameter of PSII primary photochemistry is estimated by the ratio $F_v/F_m = (F_m - F_o)/F_m$ for dark adapted leaves (Maxwell and Johnson 2000), where F_o is the minimum fluorescence, and F_m is the maximal fluorescence (Marwood et al. 2001). The effective quantum yield parameter of PSII (Φ_{PSII}) photochemistry is calculated as $\Delta F/F'_m = (F'_m - F_t)/F'_m$, where F_t is the steady-state fluorescence and F'_m is the maximum fluorescence under steady-state conditions of continuous actinic light (Maxwell and Johnson 2000). Photochemical (directly dependent on electron transport) and nonphotochemical (all non-radiative mechanisms involved in dissipation of excess absorbed light energy) quenching of fluorescence parameters can be calculated as $qP = (F'_m - F_t)/(F'_m - F'_o)$ and $qN = (F'_m - F'_o)/(F_m - F_o)$, respectively, where F'_o, is the minimum fluorescence in the light-adapted state (Genty et al. 1989; Maxwell and Johnson 2000). Because nonphotochemical quenching generally increases in plants under stress, the parameter $1 - qN$ is typically employed as an index for characterizing nonphotochemical contribution (Marwood et al. 2001). A comprehensive review of chlorophyll fluorescence is provided in Maxwell and Johnson (2000).

Most modern fluorometers use modulated measuring systems (allowing the yield of fluorescence to be measured in the presence of background illumination), with the ability to apply far-red light to plant tissue. Gas exchange measurements are often performed with an oxygen electrode, if measures are taken on submerged plants and algae (Binzer and Middelboe 2005), or with an infrared gas analyzer (IRGA), if gas exchange is measured on aerial leaves (Flexas et al. 2007).

11.3 Applications to Evaluate Xenobiotic Stress

11.3.1 Chlorophyll Fluorescence

Chlorophyll fluorescence is affected by chemicals that interfere directly with the PSII electron transport chain (mainly herbicides), or otherwise increase production of ROS that are damaging to PSII. Lipophilic compounds (e.g., PAHs, creosote, anthracene and the hydrophobic components of wastewater) have been shown to affect fluorescence parameters in aquatic plants and algae at earlier times, or at comparable or lower concentrations, than those affecting growth (Gensemer et al. 1999; Juneau et al. 2003; Marwood et al. 2001, 2003). Heavy metals including Cr, Cu, Cd and Zn, have also shown effects on chlorophyll fluorescence; only Cr, however, shows such effects at concentrations lower than those affecting growth (Appenroth et al. 2001; Drinovec et al. 2004).

Marwood et al. (2001) found that the estimated EC_{50}s for inhibition of growth, chlorophyll content and photosynthesis (both F_v/F_m and $\Delta F/F'_m$) were comparable in *L. gibba* and *M. spicatum,* exposed to creosote. Furthermore, the slopes of the concentration–response curves were similar for growth and chlorophyll-a-fluorescence endpoints, suggesting that PAH toxicity was associated primarily with effects on the photosynthetic apparatus (Marwood et al. 2001). $\Delta F/F'_m$, the quantum efficiency of electron transport through PSII, was found to be the most sensitive chlorophyll-a fluorescence parameter. Marwood et al. (2001) states that even a moderate reduction of chlorophyll-a fluorescence will cause deleterious effects on photosynthetic energy production, because of the tight relationship between these two processes; eventually the plant will suffer severely diminished growth.

The effect of herbicides that catalyze the formation of ROS (diquat and paraquat), and interfere with PSII (atrazine, bentazon, diuron, irgarol, linuron and terbuthylazine) or block the synthesis of light absorbing carotenoids can be detected earlier (and at lower concentrations) using fluorescence, than can effects on growth (Cedergreen et al. 2004; Chesworth et al. 2004; Christensen et al. 2003; Frankart et al. 2003; Hulsen et al. 2002). Clodinafop, a fatty acid inhibitor, can affect fluorescence parameters shortly after application (Abbaspoor and Streibig 2005). However, correlations with photosystem inhibition are less distinct for herbicides that disturb plant metabolism at sites more remotely connected with photosynthesis. Herbicides such as glyphosate, the imidazolinones and the sulfonylureas all block the synthesis of amino acids (Copping and Hewitt 1998), which can promote imbalances between metabolites leading to indirect effects. Plants treated with these herbicides typically take d to week to die, depending on species and environmental conditions. Changes in fluorescence parameters for these herbicides can be detected, though usually at higher concentrations, or at times when visual stress symptoms already exist (Cedergreen et al. 2004; Christensen et al. 2003; Frankart et al. 2003; Riethmuller-Haage et al. 2006).

Frankart et al. (2003) evaluated the effects of four different herbicides in *L. minor* using PAM fluorescence. For exposures to paraquat (ROS inducer in PSI), norflurazon (carotenoid biosynthetic inhibitor), and atrazine (PSII electron transport inhibitor), the percent inhibition of growth, chlorophyll and carotenoid, F_v/F_m, and F'_v/F'_m were generally similar for a given exposure concentration. However, Φ_{PSII} consistently yielded 2–3 times greater sensitivity compared to growth. In contrast, fluorescence measures were less sensitive than either growth or pigment content for exposure to flazasulfuron (branched chain amino acid inhibitor). Photosynthetic parameters reflect differing modes of action where pronounced inhibition of a parameter can be associated with a particular point in the photosynthetic process (Frankart et al. 2003). Although the results for *qN* were more variable in response and sensitivity, when compared to F_v/F_m, Φ_{PSII} and F'_v/F_m, the variables comprising qN may provide valuable information regarding xenobiotic mode of action (Frankart et al. 2003).

Growth conditions influence how early, and at which concentrations, changes in fluorescence can be measured. In greenhouse studies, effects from glyphosate have been measured after 6 and 24 hr. Comparatively, changes in fluorescence could not be measured at all after one week when exposed in growth chambers (at lower irradiance), even at concentrations above the EC_{50} (Christensen et al. 2003). If plants are growing at irradiances near photoinhibiting intensities, as occurs in greenhouses during the summer, minor herbicide-induced changes in metabolism may be sufficient to push the ROS scavenging and PSII repair systems to their limits. As a result, photosystem damage would be reflected in altered fluorescence. In comparison, well watered and fertilized plants, growing at sub-saturating irradiance, would have greater potential for up-regulating ROS scavenging and PSII repair systems. Hence, changes in fluorescence would be seen at a later time and at higher concentrations, under these growth conditions.

11.3.2 Correlations with Gas Exchange

Under laboratory conditions, there is a strong linear relationship between Φ_{PSII} and carbon fixation efficiency (Fryer et al. 1998). A linear plot of the quantum yield of CO_2 fixation (Φ_{CO2}) and PSII photochemistry (Φ_{PSII}), allows the electron requirement per molecule of CO_2 fixed to be quantified (Epron et al. 1995). However, discrepancies may occur under certain stress conditions from changes in the rate of photorespiration or pseudocyclic electron transport (Fryer et al. 1998; Genty et al. 1989; Maxwell and Johnson 2000). This correlation (Φ_{CO2} and Φ_{PSII}) may break down under field conditions, where discrepancies are caused by changes in relative rates of CO_2 fixation and competing processes such as photorespiration, nitrogen metabolism and electron donation to oxygen (Fryer et al. 1998; Maxwell and Johnson 2000).

In maize plants, comparisons of Φ_{CO2} and Φ_{PSII} showed that, when exposed to low temperatures, the ratio of Φ_{PSII}/Φ_{CO2} was higher than for unstressed fully developed leaves (Fryer et al. 1998). This was accompanied by an increase in the capacity of antioxidation systems, which indicated that leaves were suffering from oxidative stress and increased electron transport to alternative electron sinks, probably generating active oxygen species (Fryer et al. 1998).

Riethmuller-Haage et al. (2006) found that climate chamber grown black nightshade plants, treated with a sulfonylurea herbicide (metsulfuron methyl), showed similar reductions in CO_2 fixation and Φ_{PSII} (76% and 66%, respectively), compared to controls, at an illumination level of 750 μmol m^2 s^{-1}, 4 d after treatment. For greenhouse grown plants, results were similar at 70%, and 57%, respectively (Riethmuller-Haage et al. 2006). However, for glyphosate sprayed barley, gas exchange, monitored continuously for 7 d after spraying, responded for all tested concentrations within 24 hr, whereas no change in fluorescence (Kautsky-curve parameters) was detected even after 7 d (Cedergreen, personal observations).

With the exception of C_{14} incorporation or O_2 evolution, measured on micro algae, few studies use gas exchange measurements as effect measures of chemical toxicity, (Petersen and Kusk 2000). Similar to pigment content, photosynthetic rates measured on micro algae constitutes an indirect measure of algae biomass or growth, rather than an early indicator of altered plant or algae metabolism.

11.4 Advantages

Chlorophyll fluorescence is a fast, cheap, non-destructive biomarker for a large range of chemicals; effects can also be detected at an earlier stage than by measuring growth rates. Chlorophyll fluorescence has demonstrated comparable or greater sensitivity than growth endpoints for a number of contaminants, depending on the mechanism of action. Because fluorescence is a non-destructive measure, the kinetics of the toxic effect on photosynthesis can be measured over time, making it a powerful tool for assessing uptake rates, effects, internal transportation and recovery in plants (Abbaspoor et al. 2006; Cedergreen et al. 2004). The measurement of plant growth, as a toxicity endpoint, is not practical in the field, because measurements at zero time usually cannot be taken; chlorophyll fluorescence alternatively, is ideal for rapidly assessing instantaneous and/or continuous effects of contaminants (Marwood et al. 2001). Chlorophyll fluorescence depicts the efficiency of PSII, and the extent to which it is being damaged by excess light, and often provides direct measures of the photosynthetic rate (Maxwell and Johnson 2000).

11.5 Disadvantages

Fluorescence is not equally effective for all chemicals. The absence of a fluorescence response does not mean a plant is chemically un-stressed, because chlorophyll fluorescence is affected by environmental factors other than chemical stress. Hence, the most consistent results are obtained under controlled conditions, i.e., in the laboratory. If measurements are made in the field, care should be taken to interpret data in the context of other factors that can affect fluorescence. Although fluorescence is easy to measure, if experiments are not designed correctly results may be impossible to interpret (Maxwell and Johnson 2000). Practical problems with leaf darkening in the field, required for application of far-red light to plant tissue (for calculating F_0), may occur, though this can be overcome by transiently covering the leaf with a black cloth whilst simultaneously providing far-red light (Maxwell and Johnson 2000). Alternatively, clips, designed to fit the fluorescence measurement tool, are used to darken parts of leaves prior to measurements (Christensen et al. 2003).

12 Summary

This review emphasizes the predictive ability, sensitivity and specificity of aquatic plant biomarkers as biomonitoring agents of exposure and effect. Biomarkers of exposure are those that provide functional measures of exposure that are characterized at a sub-organism level. Biomarkers of effect require causal linkages between the biomarker and effects, measured at higher levels of biological organization. With the exception of pathway specific metabolites, the biomarkers assessed in this review show variable sensitivity and predictive ability that is often confounded by variations in growth conditions, rendering them unsuitable as stand alone indicators of environmental stress.

The use of gene expression for detecting pollution has been, and remains immature; this immaturity derives from inadequate knowledge on predictive ability, sensitivity and specificity. Moreover, the ability to the detect mode of action of unknown toxicants using gene expression is not as clear-cut as initially hypothesized. The principal patterns in gene expression are generally derived from stress induced genes, rather than on ones that respond to substances with known modes of action (Baerson et al. 2005). Future developments in multivariate statistics and chemometric methods that enhance pattern analyses in ways that could produce a "fingerprint", may improve methods for discovering modes of action of unknown toxicants.

Pathway specific metabolites are unambiguous, sensitive, correlate well to growth effects, and are relatively unaffected by growth conditions. These traits make them excellent biomarkers under both field and laboratory conditions. Changes in metabolites precede visible growth effects; therefore, measuring changes in metabolite concentrations reduces experimental times considerably and increases diagnostic resolution (Harring et al. 1998; Shaner et al. 2005).

The metabolic phase I enzymes (primarily associated with P-450 activity) are non-specific biomarkers, and few studies relate them to growth parameters. P-450 activity both increases and decreases in response to chemical stress, often confounding interpretation of experimental results. Alternatively, phase II metabolic enzymes (e.g., glutathione *S*-transferases; GST's) appear to be sensitive biomarkers of exposure, and potentially effect. Some GST's are affected by growth factors, but others may only be induced by xenobiotics. Measuring xenobiotic-induced GST's, or their gene expression patterns, are good candidates for future biomarkers of the cumulative load of chemical stress, both in the laboratory and under field conditions.

Phytochelatins respond to some but not all metal ions, and may therefore be used as biomarkers of exposure to identify the presence and bioavailability of ions to which they respond. However, more data on their specificity to, and interactions with growth factors, in more species are needed. The flavenoids are only represented by one heavy metal exposure study; therefore, their use as biomarkers is currently difficult to judge.

Stress proteins tend to be specific for toxicants that affect protein function. Growth factors are known to affect the level of stress proteins; hence, the use of stress proteins as biomarkers will be confined to experiments performed under controlled growth conditions, where they can be excellent indicators of proteotoxicity.

Reactive oxygen species (ROS), ROS scavenging enzymes, changes in pigment content, photosynthesis and chlorophyll fluorescence are all affected by growth factors, particularly light and nutrient availability. Therefore, these biomarkers are best suited to investigate the mode of action of toxicants under controlled growth conditions. These biomarkers are sensitive to xenobiotic stressors that affect various processes in the photosynthetic apparatus, and can be used to diagnose which photosynthetic process or processes are primarily affected. Chlorophyll fluorescence is a non-destructive measure, and is thereby well suited for repeated measures of effect and recovery (Abbaspoor and Streibig 2005; Abbaspoor et al. 2006; Cedergreen et al. 2004).

Bi-phasic responses (over time and with dose) are probably major sources of variation in sensitivity for many biomarkers. Metabolic enzymes, stress proteins, ROS and their corresponding scavenging enzymes increase in a time-frame and at doses in which plant cell damage is still repairable. However, when toxicity progresses to the point of cell damage, the concentration/activity of the biomarker either stabilizes or decreases. Examples of this response pattern are given in Lei et al. (2006); Pflugmacher et al. (2000b); Teisseire et al. (1998); and Teisseire and Guy (2000). Gene expression is also a time-dependent phenomenon varying several fold within a few hour. Therefore, bi-phasic response patterns make timing and dose-range, within which the biomarkers can be used as measures of both exposure and effect, extremely important. As a result, most biomarkers are best suited for situations in which the time and dose dependence of the biomarker, in the investigated species, are established.

Notwithstanding the previously mentioned limitations, all assessed biomarkers provide valuable information on the physiological effects of specific stressors, and are valuable tools in the search for understanding xenobiotic modes of action. However, the future use of aquatic plant biomarkers will probably be confined to laboratory studies designed to assess toxicant modes of action, until further knowledge is gained regarding the time, dose and growth-factor dependence of biomarkers, in different species. No single biomarker is viable in gaining a comprehensive understanding of xenobiotic stress. Only through the concomitant measurement of a suite of appropriate biomarkers will our diagnostic capacity be enhanced and the field of ecotoxicology, as it relates to aquatic plants, advanced.

Acknowledgments This work was supported through the Natural Science Engineering Research Council of Canada (NSERC) post-doctoral fellowship program and Baylor University.

References

Abbaspoor M, Streibig JC (2005) Clodinafop changes the chlorophyll fluorescence induction curve. Weed Sci 53:1–9.

Abbaspoor M, Teicher HB, Streibig JC (2006) The effect of root-absorbed PSII inhibitors on Kautsky curve parameters in sugar beet. Weed Res 46:226–235.

Adams SM, Giesy JP, Tremblay LA, Eason CT (2001) The use of biomarkers in ecological risk assessment: Recommendations from the Christ church conference on biomarkers in ecotoxicology. Biomarkers 6:1–6.

Akhtar TA, Lampi MA, Greenberg BM (2005) Identification of six differentially expressed genes in response to copper exposure in the aquatic plant *Lemna gibba* (Duckweed). Environ Toxicol Chem 24:1705–1715.

Amaya-Chávez A, Martínez-Tabche L, López-López E, Galar-Martínez M (2006) Methyl parathion toxicity to and removal efficiency by *Typha latifolia* in water and artificial sediments. Chemosphere 63:1124–1129.

Amrhein N, Deus B, Gehrke P, Steinröcken HC (1980) The site of the inhibition of the shikimate pathway by glyphosate: II. Interference of glyphosate with chorsimate formation *in vivo* and *in vitro*. Plant Physiol 66:830–834.

Amrhein N, Johänning D, Schab J, Schulz A (1983) Biochemical basis for glyphosate-tolerance in a bacterium and a plant tissue culture. FEBS (Fed Eur Biochem Soc) Lett 157:191–196.

Anderson JV, Davis DG (2004) Abiotic stress alters transcript profiles and activity of glutathione S-transferase, glutathione peroxidase, and glutathione reductase in *Euphorbia esula*. Physiol Plant 120:421–433.

Appenroth KJ, Stockel J, Srivastava A, Strasser RJ (2001) Multiple effects of chromate on the photosynthetic apparatus of *Spirodela polyrhiza* as probed by OJIP chlorophyll a fluorescence measurements. Environ Pollut 115:49–64.

Ahrens WH 1994 (ed.) Herbicide Handbook 7th Ed. Weed Science Society of America, Champaign, IL, 352pp.

Azevedo H, Gomes PC, Conceicao S (2005) Cadmium effects in sunflower: Membrane permiability and changes in catalase and peroxidase activity in leaves and calluses. J Plant Nutr 28:2233–2241.

Babu ST, Akhtar TA, Lampi MA, Tripuranthakam S, Dixon GD, Greenberg BM (2003) Similar stress responses are elicited by copper and ultraviolet radiation in the aquatic plant *Lemna gibba*: Implication of reactive oxygen species as common signals. Plant Cell Physiol 44:1320–1329.

Babu ST, Tripuranthakam S, Greenberg BM (2005) Biochemical responses of the aquatic higher plant *Lemna gibba* to a mixture of copper and 1,2-dihydroxyanthraquinone: Synergistic toxicity via reactive oxygen species. Environ Toxicol Chem 24:3030–3036.

Baerson SR, Sáncheez-Moreiras A, Pedrol-Bonjoch N, Schulz M, Kagan IA, Agarwal AK, Reigosa MJ (2005) Detoxification and transcriptome response in Arabidopsis seedlings exposed to the allelochemical benzoxazolin-2(3H)-one. J Biol Chem 280:21867–21881.

Barko JW (1983) Influences of light and temperature on chlorophyll composition in submersed freshwater macrophytes. Aquat Bot 15:249–255.

Beauchamp C, Fridovich I (1971) Superoxide dismutase. Improved assays and an assay applicable to acrylamide gels. Anal Biochem 44:276–287.

Behl C, Davis JB, Lesley R, Schubert D (1994) Hydrogen peroxide mediates amyloid b protein toxicity. Cell 77:817–827.

Bierkens J (2000) Applications and pitfalls of stress-proteins in biomonitoring. Toxicology 153:61–72.

Bierkens J, Maes J, Vander Plaetse F (1998) Dose-dependent induction of heat shock protein 70 synthesis in *Raphidocelis subcapitata* following exposure to different classes of environmental pollutants. Environ Pollut 101:91–97.

Binzer T, Middelboe AL (2005) From thallus to communities: Scale effects and photosynthetic performance in macroalgae communities. Marine Ecology-Progress Series 287:65–75.

Bolwell GP, Wojtaszek P (1997) Mechanisms for the generation of reactive oxygen species in plant defence—a broad perspective. Physiol Molecular Plant Physiol 51:347–366.

Bowler C, Montagu MV, Inze D (1992) Superoxide dismutase and stress tolerance. Ann Rev Plant Physiol and Plant Molecular Biol 43:83–116.

BPCP (2002) The e-Pesticide Manual Database v2.1, British Crop Protection Agency, Software developed by Wise & Loveys Information Services Ltd.

Bradford MM (1976) A rapid and sensitive method for quantitation of microgram quantities of protein utilizing the principle of protein-dye-binding. Anal Biochem 72:248–254.

Brain RA, Johnson DJ, Richards RA, Hanson ML, Sanderson H, Lam MW, Young C, Mabury SA, Sibley PK, Solomon KR (2004a) Microcosm evaluation of the effects of an eight pharmaceutical mixture to the aquatic macrophytes *Lemna gibba* and *Myriophyllum sibiricum*. Aquat Toxicol 70:23–40.

Brain RA, Johnson DJ, Richards SM, Sanderson H, Sibley PK, Solomon KR (2004b) Effects of 25 pharmaceutical compounds to *Lemna gibba* using a seven-day static renewal test. Environ Toxicol Chem 23:371–382.

Brain RA, Bestari KT, Sanderson H, Hanson ML, Wilson CJ, Johnson DJ, Sibley PK, Solomon KR (2005a) Aquatic microcosm assessment of the effects of tylosin on *Lemna gibba* and *Myriophyllum spicatum*. Environ Pollut 133:389–401.

Brain RA, Wilson CJ, Johnson DJ, Sanderson H, Bestari B-J, Hanson ML, Sibley PK, Solomon KR (2005b) Toxicity of a mixture of tetracyclines to *Lemna gibba* and *Myriophyllum sibiricum* evaluated in aquatic microcosms. Environ Pollut 138:426–443.

Brain RA, Reitsma TS, Lissemore LI, Bestari B-J, Sibley PK, Solomon KR (2006) Herbicidal effects of statin pharmaceuticals in *Lemna gibba*. Environ Sci Technol 40:5116–5123.

Buchanan BB, Gruissem W, Jones RL (2001) Photosynthesis In: Biochemistry and Molecular Biology of Plants. Buchanan BB, Gruissem W, Jones RL (eds.) American Society of Plant Physiologists, Rochville, MD, pp. 568–629.

Cedergreen N, Madsen TV (2003) Light regulation of root and leaf NO3- uptake and reduction in the floating macrophyte *Lemna minor*. New Phytol 161:449–457.

Cedergreen N, Andersen L, Olesen CF, Spliid NH, Streibig JC (2004) Does the effect of herbicide pulse exposure on aquatic plants depend on Kow or mode of action? Aquat Toxicol 71:261–271.

Cedergreen N, Abbaspoor M, Sørensen H, Streibig JC (2007) Is mixture toxicity measured on a biomarker indicative of what happens on a population level? A study on *Lemna minor*. Ecotox Environ Safe (submitted for publication).

Chesworth JC, Donkin ME, Brown MT (2004) The interactive effects of the antifouling herbicides Irgarol 1051 and Diuron on the seagrass *Zostera marina* (L.). Aquat Toxicol 66:293–305.

Christensen MG, Teicher HB, Streibig JC (2003) Linking fluorescence induction curve and biomass in herbicide screening. Pest Manag Sci 59:1303–1310.

Copping LG, Hewitt HG (1998) Chemistry and mode of action of crop protection agents. The Royal Society of Chemistry, Cambridge, UK, 160pp.

Croteau R, Kutchan TM, Lewis NG (2000) Natural products In: Biochemistry & Molecular Biology of Plants. Buchanan BB, Gruissem W, Jones RL (eds.) American Society of Plant Physiology, Rockville, MD, pp. 1250–1268.

Davy M, Petrie R, Smrchek J, Kuchnicki T, Francois D (2001) Proposal to update non-target plant toxicity testing under NAFTA, USEPA, Washington, DC, 158pp. http://www.epa.gov/scipoly/sap/.

de Pomerai DI (1996) Heat-shock proteins as biomarkers of pollution. Human Exp Toxicol 15:279–285.

Deltoro VI, Calatayud A, Morales F, Abadía A, Barreno E (1999) Changes in net photosynthesis, chlorophyll fluorescence and xanthophyll cycle interconversions during freeze-thaw cycles in the Mediterranean moss *Leucodon sciuroides*. Oecologia 120:499–505.

Deng F, Hatzios KK (2002) Purification and characterization of two glutatione S-transferase isozymes from Indica-type rice involved in herbicide detoxification. Pestic Biochem Physiol 72:10–23.

Dixon RA, Paiva NL (1995) Stress-induced phenylpropanoid metabolism. The Plant Cell 7:1085–1097.

Dixon DP, Cummins I, Cole DJ, Edwards R (1998) Glutathione-mediated detoxification systems in plans. Cur Opinion in Plant Biol 1:258–266.

Donkin P (1994) Quantitative structure-activity relationships In: Handbook of Ecotoxicology. Calow P (ed.) Blackwell Scientific Publications, Oxford, pp. 321–347.

Drinovec L, Drobne D, Jerman I, Zrimec A (2004) Delayed fluorescence of *Lemna minor*: A biomarker of the effects of copper, cadmium, and zinc. Bull Environ Contam Toxicol 72:896–902.

Edwards R, Dixon DP (2000) The role of glutathione transferases in herbicide metabolism In: Herbicides and their Mechanisms of Action. Cobb AH, Kirkwood RC (eds.), Sheffield Academic Press, Sheffield, pp. 38–71.

Eloranta P, Lahtinen T, Salonen H (1988) Effects of some environmental factors on the pigments of duckweed (*Lemna minor* L.). Aqua Fennica 18:75–84.

Epron D, Godard D, Cornic G, Genty B (1995) Limitation of net CO2 assimilation rate by internal resistances to CO2 transfer in the leaves of two tree species (*Fagus sylvatica* L.and *Castanea sativa* Mill.). Plant Cell and Environ 18:43–51.

Ernst WHO, Peterson PJ (1994) The role of biomarkers in environmental assessment (4): Terrestrial plants. Ecotoxicol 3:180–192.

Ferrat L, Pergent-Martini C, Roméo M (2003) Assessment of the use of biomarkers in aquatic plants for the evaluation of environmental quality: Application to seagrasses. Aquat Toxicol 65:187–204.

Flexas J, Fiaz-Espejo A, Berry JA, Cifre J, Galmés J, Kaldenhoff R, Medrano H, Ribas-Carbó M (2007) Analysis of leakage in IRGA's leaf chambers of open gas exchange systems: Quantification and its effects in photosynthesis parameterization. J Exp Bot 58:1533–1543.

Forbes VE, Palmqvist A, Bach L (2006) The use and misuse of biomarkers in ecotoxicology. Environ Toxicol Chem 25:272–280.

Force L, Critchley C, van Rensen JJS (2003) New fluorescence parameters for monitoring photosynthesis in plants. Photosynthesis Res 78:17–33.

Foyer CH, Lelandais M, Kunert KJ (1994) Photooxidative stress in plants. Physiol Plant 92:696–717.

Frankart C, Eullaffroy P, Vernet G (2003) Comparative effects of four herbicides on non-photochemical fluorescence quenching in *Lemna minor*. Environ Exp Bot 49:159–168.

Fryer MJ, Andrews JR, Oxborough K, Blowers DA, Baker NR (1998) Relationship between CO2 assimilation, photosynthetic electron transport, and active O2 metabolism in leaves of maize in the field during periods of low temperature. Plant Physiol 116:571–580.

Gensemer RW, Dixon DG, Greenberg BM (1999) Using chlorophyll a fluorescence to detect the onset of anthracene photoinduced toxicity in *Lemna gibba*, and the mitigating effects of a commercial humic acid. Limnol Oceanogr 44:878–888.

Genty B, Briantais J-M, Baker NR (1989) The relationship between the quantum yield of photosynthetic electron transport and quenching of chlorophyll fluorescence. Biochim Biophys Acta 990:87–92.

Greenberg BM, Huang X-D, Dixon DG (1992) Applications of the aquatic higher plant *Lemna gibba* for ecotoxicological risk assessment. J Aquat Ecosys Health 1:147–155.

Grill E, Winnacker E-L, Zenk MH (1987) Phytochelatins, a class of heavy-metal-binding peptides from plants, are functionally analogous to metallothioneins. Proc Natl Acad Sci USA 84:439–443.

Grill E, Lüffler S, Winnacker E-L, Zenk MH (1989) Phytochelatins, the heavy-metal-binding peptides of plants, are synthesized from glutathione by a specific -glutamylcysteine dipeptidyl transpeptidase (phytochelatin synthase). Proc Natl Acad Sci USA 86:6838–6842.

Gupta M, Rai UN, Tripathi RD, Chandra P (1995) Lead induced changes in glutathione and phytochelatin in *Hydrilla Verticillata* (L.f.) Royle. Chemosphere 30:2011–2020.

Hanson ML, Sibley PK, Ellis DA, Mabury SA, Muir DCG, Solomon KR (2002) Evaluation of monochloroacetic acid (MCA) degradation and toxicity to *Lemna gibba*, *Myriophyllum spicatum*, and *Myriophyllum sibiricum* in aquatic microcosms. Aquat Toxicol 61:251–273.

Hanson ML, Sibley PK, Mabury SA, Muir DCG, Solomon KR (2003) Field level evaluation and risk assessment of the toxicity of dichloroacetic acid to the aquatic macrophytes *Lemna gibba*, *Myriophyllum spicatum*, and *Myriophyllum sibiricum*. Ecotox Environ Safe 55:46–63.

Harring T, Streibig JC, Husted S (1998) Accumulation of shikimic acid: A technique for screening glyphosate efficacy. J Agric Food Chem 46:4406–4412.

Hulsen K, Minne V, Lootens P, Vandecasteele P, Hüfte M (2002) A chlorophyll a fluorescence-based *Lemna minor* bioassay to monitor microbial degredation of nanomolar concentrations of linuron. Environ Microbiol 4:327–337.

Ireland EH, Harding SJ, Bonwick GA, Jones M, Smith CJ, Williams JHH (2004) Evaluation of heat shock protein 70 as a biomarker of environmental stress in *Fucus serratus* and *Lemna minor*. Biomarkers 9:139–155.

Iwashina T (2000) The structure and distribution of the flavonoids in plants. J Plant Res 113:287–299.

Jamers A, Van der Ven K, Moens L, Robbens J, Potters G, Guisez Y, Blust R, De Coen W (2006) Effect of copper exposure on gene expression profiles in *Chlamydomonas reinhardtii* based on microarray analysis. Aquat Toxicol 80:249–260.

Jansen MAK, Babu TS, Heller D, Gaba V, Mattoo AK, Edelman M (1996) Ultraviolet-B effects on *Spirodela oligorrhiza*: Induction of different protection mechanisms. Plant Sci 115:217–223.

Jansen MAK, Hill LM, Thorneley RNF (2004) A novel stress-acclimation response in *Spirodela punctata* (Lemnaceae): 2,4,6-trichlorophenol triggers an increase in the level of an extracellular peroxidase, capable of the oxidative dechlorination of this xenobiotic pollutant. Plant Cell Environ 27:603–613.

Jensen RA (1985) The shikimate arogenate pathway: Link between carbohydrate metabolism and secondary metabolism. Physiol Plant 66:164–168.

Jin ES, Feth B, Melis A (2003) A mutant of the green alga *Dunaliella salina* constitutively accumulates zeaxanthin under all growth conditions. Biotechnol and Bioengineer 81:115–124.

Juneau P, Sumitomo H, Matsui S, Itoh S, Sang-Gil K, Popovic R (2003) Use of chlorophyll fluorescence of *Clostridium ehrenbergii* and *Lemna gibba* for toxic effect evaluation of sewage treatment plant effluent and its hydrophobic components. Ecotox Environ Safe 55:1–8.

Kautsky H, Appel W, Amann H (1960) Chlorophyllfluorescenz und kohlensaureassimilation. Biochemische Zeitschrift 322:277–292.

Kellomäki S, Wang K-Y (1999) Effects of elevated O3 and CO2 on chlorophyll fluorescence and gas exchange in scots pine during the third growing season. Environ Pollut 97:17–27.

Kidd PS, Llugany M, Poschenrieder C, Gunsé B, Barceló J (2001) The role of root exudates in aluminum resistance and silicon-induced amelioration of aluminum toxicity in three variety of maize (*Zea mays* L.). J Exp Biol 52:1339–1352.

Krebs RA, Loeschcke V (1994) Cost and benefits of activation of the heat shock response in *Drosophila melanogaster*. Funct Ecol 8:730–737.

Kreuz K, Tommasini R, Martinoia E (1996) Old enzymes for a new job. Plant Physiol 111:349–353.

Kreuz K, Martinoia E (1999) Herbicide metabolism in plants: Integrated pathways of detoxification In: The Proceedings of the 9th International Congress on Pesticide Chemistry: The Food-Environment Challenge. Brooks GT, Roberts TR (eds.) The Royal Society of Chemistry, London, pp. 279–287.

Kushwaha S, Bhowmik PC (1999) Inhibition of pigment biosynthesis in cucumber cotyledons by isoxaflutole. Photosynthetica (Prague) 37:553–558.

Lagriffoul A, Mocquot B, Mench M, Vangronsveld J (1998) Cadmium toxicity effects on growth, mineral and chlorophyll contents, and activities of stress related enzymes in young maize plants (*Zea mays* L.). Plant and Soil 200:241–250.

Lei AP, Hu ZL, Wong YS, Tam NF (2006) Antioxidant responses of microalgal species to pyrene. J Appl Phycology 18:67–78.

Lettieri T (2006) Recent applications of DNA microarray technology to toxicology and ecotoxicology. Environ Health Perspect 114:4–9.

Lewis S, Handy RD, Cordi B, Billinghurst Z, Depledge MH (1999) Stress proteins (HSP's): Methods of detection and their use as an environmental biomarker. Ecotoxicol 8:351–368.

Lewis S, Donkin ME, Depledge MH (2001) Hsp70 expression in *Enteromorpha intestinalis* (Chlorophyta) exposed to environmental stressors. Aquat Toxicol 51:277–291.

Lichtenthaler HK (1987) Chlorophylls and carotenoids: Pigments of photosynthetic membranes In: Plant Cell Membranes. Packer L, Douce R (eds.) Academic Press, Cambridge, pp. 350–383.

Loper BR, Cobb WT, Anderson KA (2002) Chemical marker for ALS-inhibitor herbicides: 2-aminobutyric acid proportional in sub-lethal applications. J Agric Food Chem 50:2601–2606.

Lydon J, Duke SO (1988) Glyphosate induction of elevated levels of hydroxybenzoic acids in higher plants. J Agric Food Chem 36:813–818.

MacFarlane GR (2002) Leaf biochemical parameters in *Avicennia marina* (Forsk.) Vierh as potential biomarkers of heavy metal stress in estuarine ecosystems. Mar Poll Bull 44:244–256.

MacFarlane GR, Burchett MD (2002) Toxicity, growth and accumulation relationships of copper, lead and zinc in the grey mangrove *Avicennia marina* (Forsk.) Vierh. Mar Environ Res 54:65–84.

MacFarlane GR (2003) Chlorophyll a fluorescence as a potential biomarker of zinc stress in the grey mangrove, *Avicennia marina* (Forsk.) Vierh. Bull Environ Contam Toxicol 70:90–96.

Mahan JR, McMichael BL, Wanjura DF (1995) Methods for reducing the adverse effects of temperature stress on plants: A review. Environ Exp Bot 35:251–258.

Marinho MM, Rodrigues SV (2003) Phytoplankton of an eutrophic tropical reservoir: Comparison of biomass estimated from counts with chlorophyll-a biomass from HPLC measurements. Hydrobiol 505:77–88.

Marshall HL, Geider RJ, Flynn KJ (2000) A mechanistic model of photoinhibition. New Phytol 145:347–359.

Marwood CA, Solomon KR, Greenberg BM (2001) Chlorophyll fluorescence as a bioindicator of effects on growth in aquatic macrophytes from mixtures of polycyclic aromatic hydrocarbons. Environ Toxicol Chem 20:890–898.

Marwood CA, Bestari KJ, Gensemer RW, Solomon KR, Greenberg BM (2003) Creosote toxicity to photosynthesis and plant growth in aquatic microcosms. Environ Toxicol Chem 22:1075–1085.

Mauch F, Dudler R (1993) Differential induction of distinct glutathione-S-transferases of wheat by xenobiotics and by pathogen attack. Plant Physiol 102:1193–1201.

Maxwell K, Johnson GN (2000) Chlorophyll fluorescence—A practical guide. J Exp Bot 51:659–668.

Maya-Chavez A, Martinez-Tabche L, Lopez-Lopez E, Galar-Martinez M (2006) Methyl parathion toxicity to and removal efficiency by *Typha latifolia* in water and artificial sediments. Chemosphere 63:1124–1129.

McCarty LS, Munkittrick KR (1996) Environmental biomarkers in aquatic toxicology: Fiction, fantasy, or functional? Human Ecol Risk Assess 2:268–274.

Meyer U, Schweim P, Fracella F, Rensing L (1995) Close correlation between heat-shock response and cytotoxicity in *Neurospora crassa* treated with aliphatic alcohols and phenols. Appl Environ Microbiol 61:979–984.

Michael GL, Hertog PH, Hollman P, Dini P (1992) Optimization of a quantitative HPLC determination of potentially anticarcinogenic flavonoids in vegetables and fruits. J Agric Food Chem 40: 1591–1598.

Miean KH, Mohamed S (2001) Flavonoid (myricetin, quercetin, kaempferol, luteolin, and apigenin) content of edible tropical plants. J Agric Food Chem 49:3106–3112.

Mitrovic SM, Pflugmacher S, JamesKJ, Furey A (2004) Anatoxin-a elicits an increase in peroxidase and glutathione S-transferase activity in aquatic plants. Aquat Toxicol 68:185–192.

Mitsou K, Koulianou A, Lambropoulou D, Pappas P, Albanis T, Lekka M (2006) Growth rate effects, responses of antioxidant enzymes and metabolic fate of the herbicide propanil in the aquatic plant *Lemna minor*. Chemosphere 62:275–284.

Mocquot B, Vangronsveld J, Clijsters H, Mench M (1996) Copper toxicity in young maize (*Zea mays* L) plants: Effects on growth, mineral and chlorophyll contents, and enzyme activities. Plant and Soil 182:287–300.

Murchie EH, Horton P (1997) Acclimation to photosynthesis to irradiance and spectral quality in British plant species: Chlorophyll content, photosynthetic capacity and habitat preferences. Plant Cell and Environ 20:438–448.

Murchie EH, Horton P (1998) Contrasting patterns of photosynthetic acclimation to the light environment are dependent on the differential expression of the responses to altered irradiance and spectral quality. Plant Cell and Environ 21:139–148.

Nakano Y, Asada K (1981) Hydrogen peroxide is scavenged by ascorbate-specific peroxidase in spinach chloroplast. Plant Cell Physiol 22:867–880.

Nelson DR, Schuler MA, Paquette SM, Werck-Reichhart D, Bak S (2004) Comparative genomics of rice and *Arabidopsis*. Analysis of 727 cytochrome P450 genes and pseudogenes from a monocot and a dicot. Plant Physiol 135:756–772.

Niehl A, Lacomme C, Erban A, Kopka J, Kramer U, Fisahn J (2006) Systemic Potato virus X infection induces defence gene expression and accumulation of beta-phenylethylamine-alkaloids in potato. Functional Plant Biology 33:593–604.

Nielsen SL (1993) A comparison of aerial and submerged photosynthesis in some Danish amphibious plants. Aquat Bot 45:27–40.

Nimptsch J, Wunderlin DA, Dollan A, Pflugmacher S (2005) Antioxidant and biotransformation enzymes in *Myriophyllum quitense* as biomarkers of heavy metal exposure and eutrophication in Suquia River basin (Cordoba, Argentina). Chemosphere 61:147–157.

Nimptsch J, Pflugmacher S (2007) Ammonia triggers the promotion of oxidative stresses in the aquatic macrophyte *Myriophyllum mattogrossense*. Chemosphere 66:708–714.

Nuwaysir EF, Bittner M, Trent J, Barrett JC, Afshari CA (1999) Microarrays and toxicology: The advent of toxicogenomics. Molecular Carcinogenesis 24:153–159.

Nygaard V, Hovig E (2006) Options available for profiling small samples: A review of sample amplification technology when combined with microarray profiling. Nucleic Acids Research 34:996–1014.

Oxborough K (2004) Imaging of chlorophyll a fluorescence: Theoretical and practical aspects of an emerging technique for the monitoring of photosynthetic performance. J Exp Bot 55:1195–1205.

Paquette SM, Bak S, Feyereisen R (2000) Intron-exon organization and phylogeny in a large superfamily, the paralogous cytochrome P450 genes of *Arabidopsis thaliana*. DNA and Cell Biol 19:307–317.

Park S, Polle JEW, Melis A, Lee TK, Jin E (2006) Up-regulation of photoprotection and PSII-repair gene expression by irradiance in the unicellular green alga *Dunaliella salina*. Marine Biotechnol 8:120–128.

Paskova V, Hilscherova K, Feldmannova M, Blaha L (2006) Toxic effects and oxidative stress in higher plants exposed to polycyclic aromatic hydrocarbons and their N-heterocyclic derivatives. Environ Toxicol Chem 25:3238–3245.

Pawlik-Skowrońska B (2001) Phytochelatin production in freshwater algae *Stigeoclonium* in response to heavy metals contained in mining water; effects of some environmental factors. Aquat Toxicol 52:241–249.

Petersen S, Kusk KO (2000) Photosynthesis tests as an alternative to growth tests for hazard assessment of toxicant. Arch Environ Contam Toxicol 38:152–157.

Pflugmacher S (2004) Promotion of oxidative stress in the aquatic macrophyte *Ceratophyllum demersum* during biotransformation of the cyanobacterial toxin microcystin-LR. Aquat Toxicol 70:169–178.

Pflugmacher S, Schroder P (1995) Glutathione s-transferases in trees—Inducibility by various organic xenobiotics. Zeitschrift fur Pflanzenernahrung und Bodenkunde 158:71–73.

Pflugmacher S, Steinberg C (1997) Activity of phase I and phase II detoxication enzymes in aquatic macrophytes. J Appl Bot-Angewandte Botanik 71:144–146.

Pflugmacher S, Geissler K, Steinberg C (1999) Activity of phase I and phase II detoxication enzymes in different cormus parts of *Phragmites australis*. Ecotox Environ Safe 42:62–66.

Pflugmacher S, Schrüder P, Sandermann H (2000a) Taxonomic distribution of glutathione S-transferases acting on xenobiotics. Phytochemistry 54:267–273.

Pflugmacher S, Schwarz S, Pachur HJ, Steinberg CEW (2000b) Effects of tributyltin chloride (TBTCl) on detoxication enzymes in aquatic plants. Environ Toxicol 15:225–233.

Pflugmacher S, Wiegand C, Beattie KA, Krause E, Steinberg CEW, Codd GA (2001) Uptake, effects, and metabolism of cyanobacterial toxins in the emergent reed plant *Phragmites australis* (cav.) trin. ex steud. Environ Toxicol Chem 20:846–852.

Porra RJ (2002) The chequered history of the development and use of simultaneous equations for the accurate determination of chlorophylls a and b. Photosyn Res 73:149–156.

Porra RJ, Thompson WA, Kriedemann PE (1989) Determination of accurate extinction coefficients and simultaneous equations for assaying chlorophylls a and b extracted with four different solvents: Verification of the concentration of chlorophyll standards by atomic absorption spectroscopy. Biochim Biophys Acta 975:384–394.

Powell RL, Kimerle RA, Moser M (1996) Development of a plant bioassay to assess toxicity of chemical stressors to emergent macrophytes. Environ Toxicol Chem 15:1570–1576.

Qi XM, Li PJ, Liu W, Xie LJ (2006) Multiple biomarkers response in maize (*Zea mays* L.) during exposure to copper. J Environ Sciences-China 18:1182–1188.

Ray TB (1984) Site of action of chlorsulfuron: Inhibition of valine and isoleucine biosynthesis in plants. Plant Physiol 75:827–831.

Reichhart D, Salaön J-P, Benveniste I, Durst F (1979) Induction by manganese, ethanol phenobarbital, and herbicides of microsomal cytochrome P-450 in higher plant tissues. Arch Biochem Biophys 196:301–303.

Reichhart D, Salaön J-P, Benveniste I, Durst F (1980) Time course of induction of cytochrome P-450, NADPH-cytochrome c reductase, and cinnamic acid hydroxylase by phenobarbital, ethanol, herbicides, and manganese in higher plant microsomes. Plant Physiol 66:600–604.

Riethmuller-Haage I, Bastiaans L, Kropff MJ, Harbinson J, Kempenaar C (2006) Can photosynthesis-related parameters be used to establish the activity of acetolactate synthase-inhibiting herbicides on weeds? Weed Sci 54:974–982.

Ritchie RJ (2006) Consistent sets of spectrophotometric chlorophyll equations for acetone, methanol and ethanol solvents. Photosyn Res 89:27–41.

Ritossa F (1962) A new puffing pattern induced by heat shock and DNP in *Drosophila*. Experientia (Basel) 18:571–573.

Roger MJR, Weiss O (2001) Fluorescence techniques In: Handbook of Plant Ecophysiology Techniques. Roger MJR (ed.) Kluwer Academic Publishers, Dordrecht, Germany, pp. 155–172.

Rowan KS (1989) Photosynthetic pigments of algae. Cambridge University Press, Cambridge, 334pp.

Roy S, Hanninen O (1994) Pentachlorophenol—Uptake elimination kinetics and metabolism in an aquatic plant, *Eichhornia crassipes*. Environ Toxicol Chem 13:763–773.

Roy S, Lindstrüm-Seppä P, Huuskonen S, Hänninen O (1995) Responses of biotransformation and antioxidant enzymes in *Lemna minor* and *Oncorhynchus mykiss* exposed simultaneously to hexachlorobenzene. Chemosphere 30:1489–1498.

Sánchez-Estudillo L, Freile-Pelegrin Y, Rivera-Madrid R, Robledo D, Narva'ez-Zapata JA (2006) Regulation of two photosynthetic pigment-related genes during stress-induced pigment formation in the green alga, *Dunaliella salina*. Biotechnol Lett 28:787–791.

Sánchez-Machado DI, Lopez-Hernández J, Paseiro-Losada P, Lopez-Cervantes J (2004) An HPLC method for quantification of sterols in edible seaweeds. Biomed Chromatogr 18:183–190.

Sanders B (1990) Stress proteins: Potential as multitiered biomarkers In: Biomarkers of Environmental Contamination. McCarthy JF, Shugart LR (eds.) Lewis, Boca Raton, FL, pp. 165–191.

Sanders BM (1993) Stress proteins in aquatic organisms: An environmental perspective. Crit Rev Toxicol 23:49–75.

Sanders BM, Martin LS, Nakagawa PA, Hunter DA, Miller S, Ullrich SJ (1994) Specific cross-reactivity of antibodies raised against two major stress proteins, stress 70 and chaperonin 60, in diverse species. Environ Toxicol Chem 13:1241–1249.

Santos C, Gaspar M, Caeiro A, Branco-Price C, Teixeira A, Ferreira RB (2006) Exposure of *Lemna minor* to arsenite: Expression levels of the components and intermediates of the ubiquitin/proteasome pathway. Plant Cell Physiol 47:1262–1273.

Sayed OH (2003) Chlorophyll fluorescence as a tool in cereal crop research. Photosynthetica (Prague) 41:321–330.

Schrenk C, Pflugmacher S, Bruggemann R, Sandermann H, Steinberg CEW, Kettrup A (1998) Glutathione S-transferase activity in aquatic macrophytes with emphasis on habitat dependence. Ecotox Environ Safe 40:226–233.

Schwender J, Seemann M, Lichtenthaler HK, Rohmer M (1996) Biosynthesis of isoprenoids (carotenoids, sterols, prenyl side-chains of chlorophylls and plastoquinone) via a novel pyruvate/

glyceraldehyde 3-phosphate non-mevalonate pathway in the green algae *Scenedesmus obliquus*. Biochem J 316:73–80.

Sergiev IG, Alexieva VS, Ivanov SV, Moskova II, Karanov EN (2006) The phenylurea cytokinin 4PU-30 protects maize plants against glyphosate action. Pestic Biochem Physiol 85:139–146.

Shaner DL, Nadler-Hassar T, Henry WB, Koger CH (2005) A rapid *in vivo* shikimate accumulation assay with excised leaf discs. Weed Sci 53:769–774.

Simonovicova-Olle M, Bocova B, Huttova J, Mistrik I, Tamas L (2006) The effect of aluminum ions on the activity of ascorbate-redox- and oxidative stress-related enzymes in barley root tips. Periodicum Biologorum 108:593–596.

Smith IK, Vierheller TL, Thorne CA (1988) Assay of glutathione reductase in crude tissue homogenates using 5,5 -dithiobis(2-nitrobenzoic acid). Anal Biochem 175:408–413.

Srivastava S, Mishra S, Tripathi RD, Dwevedi S, Gupta DK (2006) Copper-induced oxidative stress and responses of antioxidants and phytochelatins in *Hydrilla verticillata* (L.f.) Royle. Aquat Toxicol 80:405–415.

Strasser RJ, Srivastava A, Tsimilli-Michael M (2000) The fluorescence transient as a tool to characterise and screen photosynthetic samples In: Probing Photosynthesis: Mechanisms, Regulation and Adaptation. Yunus M, Pathre U, Mohanty P (eds.) Taylor & Francis, London, pp. 445–483.

Tausz M, Sircelj H, Grill D (2004) The glutatione system as a stress marker in plant ecophysiology: Is a stress-response concept valid? J Exp Bot 55:1955–1962.

Teisseire H, Guy V (2000) Copper-induced changes in antioxidation enzymes activities in fronds of dockweed (*Lemna minor*). Plant Sci 153:65–72.

Teisseire H, Vernet G (2001) Effects of the fungicide folpet on the activities of antioxidative enzymes in duckweed (*Lemna minor*). Pestic Biochem Physiol 69:112–117.

Teisseire H, Couderchet M, Vernet G (1998) Toxic responses and catalase activity of *Lemna minor* L. exposed to folpet, copper, and their combination. Ecotox Environ Safe 40: 194–200.

Turgut C, Fomin A (2002) Sensitivity of the rooted macrophyte *Myriophyllum aquaticum* (Vell.) Verdcourt to seventeen pesticides determined on the basis of EC50. Bull Environ Contam Toxicol 69:601–608.

Turgut C, Grezichen A, Fomin A (2003) Toxicity of sulfunulurea herbicides to the dicotyledonous macrophyte *Myriophyllum aquaticum* in a 14 day bioassay. Fresenius Environ Bull 12:619–622.

USEPA (2004) Potential Implications of Genomics for Regulatory and Risk Assessment Applications at EPA, Science Policy Council U.S. Environmental Protection Agency, Washington, DC.

Waters ER, Lee GJ, Vierling E (1996) Evolution, structure and function of small heat shock proteins in plants. J Exp Bot 47:325–338.

Wetzel RG (1983) Limnology, 2nd Ed. Saunders College Publishing, Philadelphia, PA, 767pp.

Whyte JJ, Jung RE, Schmitt CJ, Tillitt DE (2000) Ethoxyresorufin-O-deethylase (EROD) activity in fish as a biomarker of chemical exposure. Crit Rev Toxicol 30:347–570.

Wilhelm C, Volkmar P, Lohmann C, Becker A, Meyer M (1995) The HPLC-aided pigment analysis of phytoplankton cells as a powerful tool in water-quality control. J Water Supply Res and Technol-Aqua 44:132–141.

Yin L, Zhou Y, Fan X, Lu, R (2002) Induction of phytochelatins in *Lemna aequioctialis* in response to cadmium exposure. Bull Environ Contam Toxicol 68:561–568.

Zenk MH (1996) Heavy metal detoxification in higher plants—A review. Gene 179:21–30.

Human Health Effects of Methylmercury Exposure

Sergi Díez

Contents

1 Introduction 111
2 Sources and Cycling of Mercury in the Global Environment 112
3 Methylmercury 114
3.1 Pathways of Human Exposure 114
3.2 Reports of Exposure 116
3.3 Toxicokinetics 116
3.4 Toxicity and Effects on Humans 119
3.5 Risk Evaluations 120
3.6 Risks and Benefits of Fish Consumption 121
3.7 Biomarkers and Exposure Evaluation 122
4 Summary 125
References 125

1 Introduction

Mercury (Hg) has caused a variety of significant and documented adverse effects on human health and the environment throughout the world. Mercury and the compounds with which it combines are highly toxic, particularly to the developing nervous system. The toxicity mercury imposes on humans and other organisms is dependent on the chemical form, the amount, the pathway of exposure and the vulnerability of the person exposed. Human exposure to mercury may occur via a variety of pathways, including consumption of fish, occupational and household uses, dental amalgams and mercury-containing vaccines.

Mercury has received special attention because it has proven toxic effects on multiple species. In particular, methylmercury (MeHg), the most toxic form of mercury, can cause severe neurological damage to humans and wildlife (Clarkson

S. Díez
Environmental Chemistry Department, IDAEA-CSIC, Jordi Girona, 18–26, E-08034, Barcelona, Spain;
Environmental Geology Department, ICTJA-CSIC, Lluis Solé i Sabarís, s/n, E-08028 Barcelona, Spain

D.M. Whitacre (ed.) *Reviews of Environmental Contamination and Toxicology*, Vol 198 111
doi: 10.1007/978-0-387-09646-9,

et al. 2003; Grandjean et al. 1999). The primary means by which humans are exposed to mercury are through contact with dental amalgams (mercury vapor, Hg^0) and fish consumption (MeHg). Recently, increasing concern has been expressed by pediatricians regarding the safety of many vaccine preparations routinely administered to infants, which contain an ethyl mercury compound (thimerosal). Although patterns of human usage of mercury have changed over the centuries, occupational exposure to it still occurs; significant sources include exposure to Hg vapor during mining (gold, etc.) operations (Grandjean et al. 1999; Malm 1998), Hg use in the chlor-alkali industry (Calasans and Malm 1997; Montuori et al. 2006) and Hg use in dentistry (Harakeh et al. 2002; Morton et al. 2004).

Today, the general population is primarily exposed to three different forms of mercury: mercury vapors emitted by dental amalgam fillings (Goering et al. 1992; Hansen et al. 2004; Razagui and Haswell 2001), MeHg naturally bioaccumulated in fish (Bjornberg et al. 2003; Canuel et al. 2006; Hightower and Moore 2003), and an ethyl mercury compound, thimerosal, which is employed as a preservative in certain commonly used childhood vaccines (Bernard et al. 2004; Halsey 1999; Sager 2006).

2 Sources and Cycling of Mercury in the Global Environment

Mercury is a natural element in the earth's crust; it is a silver-colored, shiny, liquid metal found in a variety of chemical forms in rocks, soil, water, air, plants and animals. Mercury usually combines with other elements to form various inorganic (e.g., the mineral cinnabar, a combination of mercury and sulfur), or organic (e.g., MeHg) compounds, although Hg occasionally also occurs in its elemental, relatively pure form, as a liquid or vapor.

The global cycling of mercury begins when Hg vapor rises from land and sea surfaces. Most atmospheric mercury exists as elemental mercury vapor, a chemically stable monatomic gas, which circulates in the atmosphere for up to 1 yr; mercury vapor may be widely dispersed and may be transported thousands of miles from original points of emission. Hg vapor is oxidized in the upper atmosphere to water-soluble ionic mercury, which is returned to the earth's surface in rainwater. Wet deposition during rainfall is the primary mechanism by which mercury is transported from the atmosphere to surface waters and land. After deposition, mercury commonly is emitted back to the atmosphere, either as a gas or associated with particles, to be re-deposited, elsewhere. As Hg cycles between environmental compartments (atmosphere, land, and water), mercury undergoes a series of complex chemical and physical transformations, many of which are not completely understood. In fact, about 90% of the total Hg input to oceans is recycled to the atmosphere, and less then 10% reaches sediments. However, a small percentage (about 2%) is methylated in biota and is accumulated in the food-chain; only a small fraction of MeHg is lost to the atmosphere, mainly as highly volatile dimethyl mercury (Fitzgerald et al. 1998). Mercury accumulates most efficiently in the aquatic food web, where predatory organisms at the top trophic levels have the highest mercury concentrations. Almost all of the mercury that accumulates in fish tissue is MeHg; inorganic mercury is less efficiently absorbed, and more readily eliminated from the body than is MeHg, and it does not tend to bioaccumulate.

The sources from which mercury is released to the environment can be grouped into four categories: (1) natural sources; (2) current anthropogenic releases from mobilization of mercury impurities in raw materials; (3) current anthropogenic releases resulting from mercury used intentionally in products and processes; and (4) re-mobilization of historically-deposited anthropogenic mercury releases worldwide.

Natural sources include volcanoes, evaporation from soil and water surfaces, degradation of minerals and forest fires. Available information indicates that natural sources account for less than 50% of total releases. There are indications that current global anthropogenic emissions of mercury have resulted in deposition rates that are 1.5–3 times higher than those of pre-industrial times. In the vicinity of industrial areas, the deposition rates have increased by 2–10 times during the last 200 yr (Bergan et al. 1999; USEPA 1997). In 1994, global natural Hg emissions were estimated to be ~1,650 metric t/yr (Mason et al. 1994); in a later update, 1,400 t/yr constituted the best estimate (Lamborg et al. 2002). The Programme for Monitoring and Evaluation of the Long-Range Transmission of Air Pollutants in Europe (European Model and Evaluation Program; EMEP) estimated the global natural emission at about 2,400 metric t/yr, of which 1,320 was emitted from land and 1,100 was emitted from oceans (Bergan and Rodhe 2001). Emission inventories indicate that Asian sources account for more than 50% of the global anthropogenic emissions of total Hg (Jaffe et al. 2005). In the coming decades, it is expected that the rapid economic and industrial development in Asia will result in a significant increase in anthropogenic Hg emissions, unless drastic measures are taken to prevent it (Wong et al. 2006).

Among the more important *anthropogenic processes* that mobilize mercury impurities are the following: coal-fired power plants and coal burning for heat generation; cement production; and mining and other metallurgic activities involving the extraction and processing of minerals (production of iron and steel, zinc and gold). Important sources of anthropogenic releases that occur from the intentional extraction and use of mercury include the following: mercury mining; small-scale gold and silver mining; chlor-alkali production; breakage of fluorescent lamps, auto headlamps, manometers, thermostats, thermometers, and other instruments; dental amalgam fillings; manufacturing of products containing mercury; waste treatment and incineration of products containing mercury; landfills; and cremation.

The atmospheric residence time of elemental mercury is in the range of mon to ~1 yr. The environmental residence time for atmospheric mercury makes transport, on a hemispherical scale, possible; emissions on any continent may, therefore, contribute to the deposition on other continents. For example, based on modeling of the intercontinental mercury transport performed by the EMEP at the Meteorological Synthesizing Centre-East (EMEP/MSC-E) (Travnikov 2005), up to 67% of total depositions to the continent are from external anthropogenic and natural sources. Among these, ~24% are from Asian sources, and 14% from European ones.

In Europe, about 40% of annual mercury depositions result from intercontinental transport, including 15% from Asia and 5% from North America. The Arctic region has no significant local sources of mercury emission. However, about half of the mercury deposition in the Arctic results from atmospheric transport from foreign anthropogenic emission sources; the greatest contribution is from Asian (33%) and European sources (22%).

Speciation influences the transport of mercury within and between environmental compartments. Mercury adsorbed onto particles, and ionic (e.g., divalent) mercury compounds are normally deposited on land and in water, primarily near their sources of origin; in contrast, elemental mercury vapor is transported on a hemispherical or global scale portending global concern for such vapor emissions. Another concern is the so-called "polar sunrise mercury depletion incidence," a special phenomenon that has been shown to influence the deposition of mercury in Polar regions. It has also been termed "the mercury sunrise," because deposition of high amounts of mercury is taking place during the first few mon of the Polar sunrise. It appears that solar activity and the presence of ice crystals influence the atmospheric transformation of elemental gaseous mercury to divalent mercury, which is more rapidly deposited. Mercury depletion, from this effect, has now been observed in Alert, Canada (Schroeder et al. 1998), in Barrow, Alaska, USA (Lindberg et al. 2002), Svalbard, Norway (Berg et al. 2003), in Greenland (Ferrari et al. 2004), as well as in the Antarctic (Ebinghaus et al. 2002).

3 Methylmercury

3.1 Pathways of Human Exposure

As a result of the previously-described global cycling of mercury, inorganic mercury reaches the aquatic environment and is transformed to MeHg via methylation by microbial communities in aquatic sediments. Microbial conversion to MeHg is believed to be a protective mechanism, because inorganic (mercuric) mercury is more toxic to them. In aquatic systems, the microorganisms primarily responsible for methylation of mercury are the sulfate-reducing bacteria (Acha et al. 2005; King et al. 2000; Watras et al. 2005). This MeHg produced by microbial action enters the aquatic food chain. Accordingly, when human exposure occurs, it is almost exclusively from consumption of fish and marine mammals contaminated with MeHg.

The mercury concentrations that exist in various fish species generally range between 0.01 and 4 mg/kg, depending on factors such as pH and redox potential of the water, and species, age and size of the fish. Because mercury biomagnifies in the aquatic food web, fish and other higher trophic level organisms, tend to have higher levels of mercury. Hence, large predatory fish, such as king mackerel, pike, shark, swordfish, walleye, barracuda, large tuna, scabbard and marlin, as well as seals and toothed whales, contain the highest mercury concentrations. Available data indicate that mercury is present around the globe in concentrations that may adversely affect humans and wildlife. These levels have led to advisories in a number of countries (for fish, and sometimes marine mammals) to limit or avoid consumption of certain fish taken from mercury-contaminated water bodies; warnings have been issued particularly for sensitive subgroups (pregnant women and young children). Moderate consumption of fish (with low mercury levels) is not likely to result in exposures that compromise health. However, people who consume larger quantities of fish or marine mammals may face health risks from overexposure to mercury (Hightower and Moore 2003). Although MeHg naturally accumulates in fish through the food chain, consumption of farmed fish may also result in MeHg exposures. Fish-consumption advisories issued to protect human health do not usually extend to fish by-products fed to farmed animals. Future guidelines designed to decrease exposure to MeHg must address farming practices that use fish by-products (Dorea 2006). Previous studies have shown that the meat of animals fed fish meal, or other fish products, is likely to contribute to MeHg exposure (Choi and Cech 1998; Lindberg et al. 2004). Nevertheless, results from studies have shown no significant difference in MeHg levels in wild vs. farmed salmon (Easton et al. 2002; Foran et al. 2004).

The most dramatic case of severe MeHg poisoning, which resulted from fish consumption by fisherman and their families, was the Minamata Bay, Japan incident of the 1950s. The source of contamination, in this incident, was an acetaldehyde manufacturing plant in which inorganic mercury was used as a catalyst. The amount of mercury discharged to water ways from this plant, between 1932 and 1968, was large, and was estimated at 456 t. As a result of these discharges, and subsequent consumption by the local population of MeHg contaminated-fish from the polluted water ways, hundreds of people died, and thousands were affected, many permanently (Akagi et al. 1998; Harada 1995). The medical disorders associated with this epidemic became known as "Minamata disease." A similar fish-mediated epidemic of MeHg poisoning occurred in riverside villages along the Agano River in Niigata, Japan, in 1964–1965 (Tsubaki and Irukayama 1977).

Japan is not the only country, however, in which such events have occurred. Recently, a population resident in the Brazilian Amazon showed evidence of increased exposure to MeHg, which resulted from their consumption of fish contaminated by upstream gold-mining activities (Dolbec et al. 2000; Grandjean et al. 1999; Lebel et al. 1998).

Methyl- and ethyl-mercury poisonings occurred in Iraq following consumption of seed grain that had been treated with fungicides containing these alkylmercury compounds (Bakir et al. 1973). The first outbreaks were caused by ethylmercury, and occurred in 1956 and, again in 1959–1960; ~1,000 people were adversely affected. The second outbreak was caused by MeHg, and occurred in 1972. The imported mercury-treated seed grains were subsequently ground into flour that was used to prepare homemade bread throughout rural areas of Iraq. The latent period after exposure to MeHg, before toxicity appeared, contributed to this disaster; farmers fed contaminated grain to their livestock without observing any immediate effects. Such observations probably led farmers to conclude that the grain was safe for them to consume, as well. Subsequent human consumption of contaminated bread resulted in insidious and irreversible neurological symptoms. The victims experienced no ill effects during the intake period. After consumption had stopped, no neurological symptoms appeared for more than 1 mon. This relatively long latent period is a dangerous property of MeHg. In the Iraq incident, the first symptom to appear was paresthesia and later, but in a rapid sequence, more severe signs appeared such as ataxia, dysarthria and loss of vision. Unlike the long-term exposures in Japan, the epidemic of MeHg poisoning in Iraq had a short duration, although, the magnitude of the exposure was high. It was reported that as many as 50,000 persons may have been exposed, and neurological impairments in children were evident (Myers et al. 2000). Tragically, more than 6,500 individuals were hospitalized, and 459 died from consumption of Hg-contaminated bread.

3.2 Reports of Exposure

Estimation of the magnitude of MeHg exposure to the general population has only been made in few countries such as the United States (Mahaffey et al. 2004; McDowell et al. 2004), Japan (Yasutake et al. 2004) and Germany (Becker et al. 2002, 2003). Apart from those populations known to have high fish consumption, mean Hg levels in hair generally range from >0.1 to 1.0 mg/kg (Bjornberg et al. 2003; Knobeloch et al. 2005; Montuori et al. 2006; Stern et al. 2001); mean blood levels of Hg are generally in the range of 1.0–5.0 μg/L, although, worldwide, there are more Hg residue data for hair, than for blood.

Higher levels of Hg exposure occur in populations that live near oceans, lakes and rivers, because these populations consume more freshwater fish. In the Faroe and Seychelles islands, the median Hg concentration found in maternal hair is 4.5 mg/kg (Grandjean et al. 1992), and 5.8 mg/kg (Cernichiari et al. 1995a), respectively. In the river basins of the Amazon, median hair Hg levels typically range between 5 and 15 mg/kg (Akagi et al. 1995; Barbosa et al. 2001; de Campos et al. 2002; Dolbec et al. 2001; Dorea et al. 2003; Kehrig et al. 1997; Santos et al. 2002a). Table 1 presents the Hg concentrations found in human hair for residents of various countries. It can be seen from these results, that gold mining, together with chlor-alkali plants, constitute the most important sources of mercury contamination that affect people, in the world.

3.3 Toxicokinetics

Several recent reviews have addressed the absorption, disposition, and excretion of MeHg in the body (ATSDR 1999; Clarkson and Magos 2006; USEPA 2001). It is well known that about 95% of MeHg ingested in fish is absorbed in the

Table 1 Comparison of mercury (Hg) concentrations (mg/kg dry wt) in human hair collected at various worldwide locations

Location	Mean	Range	Remarks	References
Faroe Islands	4.27	2.6–7.7	Mother at parturition	Grandjean et al.1997
	1.12	0.69–1.88	Child, 12 mon	
	2.99	1.7–6.1	Child, 7 yr	
Minamata, Japan		2.46–705	Patients of Minamata disease	Harada 1995
Tokyo	2.98		Male	Nakagawa 1995
	2.02		Female	
Wau-Bulolo, Papua, New Guinea	0.55	0.19–1.1	Background	Saeki et al. 1996
	1.20	0.39–3.0	Gold-mining area	
Bangladesh	0.44	0.02–0.95	Fish consumption	Holsbeek et al. 1996
Tarragona, Spain	0.77	0.18–2.44	School children	Batista et al. 1996
California, USA	0.64	0.3–1.8	Tribal members	Harnly et al. 1997
	1.60	0.3–2.3	Non-tribal members	
Shiranui Bay, Japan	5.00		Male	Harada et al. 1998
	2.10		Female	
Tokushima, Japan	4.62	0.626–24.6		Feng et al. 1998
Harbin, China	1.69	0.112–36.4		
Medan, Indonesia	3.13	0.203–19.9		
Hong Kong, China	3.33		Fertile male	Dickman et al. 1998
	4.23		Subfertile male	
New Zealand		3–6	Maternal hair	Crump et al. 1998
Madeira, Portugal	10.39	1.93–42.61	Pregnant woman	Renzoni et al. 1998
Lake Victoria, Tanzania		0.29–953	Gold mine	Harada et al. 1999
		0.29–416	Fishing village	
		0.48–474	City	
Seoul, Korea	1.70		Male	Lee et al. 2000
	1.10		Female	
Doha, Kuwait	4.18		Fisherman	Al-Majed and Preston 2000
	2.62		Control	
Gdansk, Poland	0.38		Person who died suddenly	Hac et al. 2000
Montreal, Canada	0.82	<0.20–6.59	Frequent fish consumers	Kosatsky et al. 2000
	0.38	<0.20–3.38	Infrequent fish consumers	
Seychelles Islands	6.80	0.5–26.7	Maternal hair	Myers et al. 2000
	6.5	0.9–25.8	5–6-yr-old children	
Diwalwal, Philippines	2.65	0.03–34.71	Control	Drasch et al. 2001
	2.77	0.03–13.17	Downstream of gold mine	
	1.71	0.03–8.91	Gold mine, non-occupational	
	3.62	0.03–37.76	Gold mine, occupational	
Kamuango, Kenya	1.44	0.67–3.5	Gold mine	Harada et al. 2001
Sori Beach, Kenya	2.09	0.73–5.6	Fishing village	
Homa Bay, Kenya	4.50	0.61–42.8	Fishing village, with Hg-containing soap user	
Dunga Beach, Kenya	48.50	0.27–900	Fishing village, with Hg-containing soap user	
Kisumu, Kenya	145	1.1–603	City, with Hg-containing soap user	
Tapajos River, Brazil		1.8–53.8	Gold-mining area	
Negro river basin, Brazil	12.65	0–44.53	<15-yr-old children	Barbosa et al. 2001
Wayana, French Guiana	11.40			Frery et al. 2001
Upper Maroni, French Guiana	12.7 (10.2)		Maternal (children) hair	Cordier et al. 2002
Camopi, French Guiana	6.7 (6.5)			
Awala, French Guiana	2.8 (1.4)			
Mansoura, Egypt	0.23	0.11–0.41	Urban area	Mortada et al. 2002
Anwiaso, Ghana	1.61	0.15–5.86	Children in gold-mining area	Adimado and Baah 2002

(continued)

Table 1 (continued)

Location	Mean	Range	Remarks	References
Sahuma, Ghana	0.62	0.32–2.19	Children in gold-mining area	
Tanoso, Ghana	4.27	0.06–28.3	Children in gold-mining area	
Elubo, Ghana	1.21	0.07–3.19	Children in gold-mining area	
Camito, Colombia	4.91		Fisherman	Olivero et al. 2002
Sai Cinza, Brazil	16.00	4.50–90.4	Gold-mining area	Santos et al. 2002b
Santana do Ituqyi, Brazil	4.33	0.40–11.60		
Aldeia do Lago Grande, Brazil	3.98	0.40–11.76		
Tabatinga, Brazil	5.37	0.37–16.96		
Caxiuana, Brazil	8.58	0.61–45.59		
Madeira Island	4.09	0.38–25.95	7-yr-old children	Murata et al. 2002
Germany	0.23	0.06–1.7	8–10-yr-old children	Pesch et al. 2002
Minamata, Japan	1.76	0.09–10.56		Yasutake et al. 2003
Kumamoto, Japan	1.57	0.14–19.18		
Tottori, Japan	2.04	0.00–12.52		
Wakayama, Japan	2.04	0.00–20.66		
Chiba, Japan	3.37	0.14–26.76		
Japan	1.64	0.45–6.32	7-yr-old children	Murata et al. 2004
Japan	2.42		Male	Yasutake et al. 2004
	1.37		Female	
USA	0.22	0.18–0.25	1–5-yr-old children	McDowell et al. 2004
	0.47	0.35–0.58	16–49-yr-old women	
Cambodia	3.10	0.54–190		Agusa et al. 2005
Kayabi, eastern Amazonia, Brazil	16.55		Children fish consumers in gold-mining area	Dorea et al. 2005
Cururu, eastern Amazonia, Brazil	4.76		Children fish consumers in gold-mining area	
Kaburua, eastern Amazonia, Brazil	2.87		Children fish consumers in gold-mining area	
Spain	0.94	0.19–5.63	4-yr-old children	Montuori et al. 2006
Menorca, Spain	0.72	0.23–3.83	4-yr-old children, diffuse source	
Flix, Spain	1.26	0.19–5.63	4-yr-old children, chlor-alkali source	
Mediterranean coast, Morocco	1.79	0.22–9.56	Frequent fish consumers	Elhamri et al. 2007
Zhoushan City, China	1.25	0.93–1.69	Maternal urban	Gao et al. 2007
Naples, Italy	0.64	0.22–3.40	Urban area	Díez et al. 2008a
Madrid and Sabadell, Spain	1.68	0.13–8.43	Neonates	Díez et al. 2008b

gastrointestinal tract, although the exact site of absorption is unknown. After consumption, MeHg is accumulated in liver and kidney, but is distributed to all tissues, a process that takes some 30–40 hr. Approximately 5% of MeHg is found in the blood and 10% in brain; in the blood, most mercury is bound to the hemoglobin in red cells, with a concentration of about 20 times the plasma concentration. The concentration in brain is about 5 times, and in scalp hair about 250 times the corresponding concentration in blood. Hair levels closely follow blood levels. MeHg crosses the blood brain and placental barriers. Moreover, levels in cord blood are proportional to, but slightly higher than, levels in maternal blood. Levels in the fetal brain are about five to seven times higher than levels in maternal blood (Cernichiari et al. 1995a).

MeHg is slowly metabolized to inorganic mercury by microflora in the intestines; the biochemical mechanisms that accomplish this are not known. Although MeHg is the predominant form of mercury at the time of exposure, inorganic mercury slowly accumulates in the body and resides for long periods in the central nervous system. Inorganic mercury is believed to be in an inert form, perhaps existing as insoluble mercury selenide (Clarkson et al. 2003).

The half-life of MeHg in the body is about 50 d, with a range of 20–70 d; the half-life in hair averages about 65 d, with a range of about 35–100 d (Clarkson 1993), indicating that MeHg leaves the body slowly. Urinary excretion is negligible: 10% or less of total elimination from the body. Most MeHg is eliminated from the body by demethylation, and excretion of the inorganic form in feces. Biliary excretion and demethylation by microflora do not occur in suckling animals. The failure of neonates to excrete MeHg may be associated with the inability of suckling infants to perform these two metabolic processes (Ballatori and Clarkson 1982; Rowland et al. 1977).

The high mobility of MeHg in the body does not result from lipid solubility, as claimed in some textbooks; rather, it results from the formation of small molecular weight thiol complexes that are readily transported across cell membranes. MeHg is present in the body as water-soluble complexes, mainly, if not exclusively, attached to the sulfur atom of thiol ligands. It enters the endothelial cells of the blood–brain barrier as a complex with L-cysteine. The attachment of MeHg to the thiol ligand in the amino acid cysteine results in a complex, the structure of which mimics that of the large neutral amino acid, L-methionine. The process is so specific, that the complex with the optical isomer D-cysteine is not transported. Thus, MeHg transport into tissues appears to be mediated by the formation of a MeHg-cysteine complex. This L-complex is structurally similar to methionine, and is transported into cells via a widely distributed neutral amino acid carrier protein (Kajiwara et al. 1996, 1997; Kerper et al. 1992). Although MeHg is distributed throughout the organs of the body, it has its most devastating effect on the developing brain.

3.4 *Toxicity and Effects on Humans*

Once MeHg is dispersed throughout the body by blood flow, and enters the brain, it may cause structural damage. The critical target for MeHg toxicity is the central nervous system. Physical lesions may be manifested as tingling and numbness in fingers and toes, loss of coordination, difficulty in walking, generalized weakness, impairment of hearing and vision, tremors, and finally loss of consciousness leading to death. The developing fetus may be at particular risk from

MeHg exposure. Infants, born to mothers exposed to MeHg during pregnancy, have exhibited a variety of developmental neurological abnormalities, including the following: delayed onset of walking and talking, cerebral palsy, altered muscle tone and deep tendon reflexes, and reduced neurological test scores. Maternal toxicity may or may not have been present during pregnancy for those offspring to exhibit adverse effects. The critical effects observed following MeHg exposure to the general population are multiple central nervous system effects including ataxia and paresthesia.

3.5 *Risk Evaluations*

The general population is primarily exposed to MeHg through the diet, especially fish and other seafood. Authorities in several countries and international organizations have used risk evaluation tools to establish prospective safe levels (RfD: Reference Dose; PTWI: Provisional Tolerable Weekly Intake), limits, advisories or guidelines for consumption of fish or other foods contaminated by mercury compounds. The RfD is defined by the United States Environmental Protection Agency (US EPA) as an estimate (with an uncertainty of approximately tenfold) of the daily exposure to humans (including sensitive subgroups) that is likely to be without an appreciable risk of deleterious effects during an entire lifetime. In contrast, the PTWI represents the maximum acceptable level of a contaminant in the diet; the goal should be to limit exposure to the extent feasible, consistent with the PTWI.

Some studies have demonstrated the neurological injuries caused by low-level MeHg exposure (Mendola et al. 2002; Grandjean et al. 2004). Recent prospective epidemiologic studies from the Faroe Islands, the Seychelles Islands and New Zealand have reported on the developmental effects of low-level maternal and fetal MeHg exposure in fish-consuming populations (Cernichiari et al. 1995b; Grandjean et al. 1997; Crump et al. 1998).

The US EPA relied on the Faroe Islands study (USEPA 1997) to establish a benchmark dose that was converted into a maternal intake of 1.1 μg Hg/kg body weight (bw) per d. After applying a safety factor of 10, an RfD of 0.1 μg/kg bw/d was recommended. The reference dose will be exceeded if a substantial amount of mercury contaminated fish is ingested. For example, if the weekly intake of fish (having residues >0.4 mg/kg) is about 100 g (one typical fish meal per week), the RfD will be exceeded. Therefore, fish mercury levels should be kept below this limit.

The 67th meeting of the Joint FAO/ WHO Expert Committee on Food Additives (JECFA) was held in 2006 to evaluate certain food additives and contaminants. During this meeting a PTWI of 1.6 μg MeHg/kg bw/week (equal to 0.23 μg MeHg/kg bw/d) for the general population (JECFA 2006) was confirmed. The JECFA established this PTWI using the most sensitive toxicological end-point, i.e., developmental neurotoxicity, in the most susceptible species (humans). However, the

JECFA noted that life-stages other than the embryo and fetus may be less sensitive to the adverse effects of MeHg.

The risks from mercury in fish and shellfish depend on the amount of fish and shellfish eaten and the levels of mercury they contain. Therefore, the US Food and Drug Administration (FDA) and the EPA are advising women who may become pregnant, pregnant women, nursing mothers, and young children to avoid some types of fish and eat fish and shellfish that are lower in mercury. Both recommend that these sensitive individuals avoid eating shark, swordfish, king mackerel, or tilefish, because they contain high levels of mercury; in addition, these sensitive groups should not eat more than 12 ounces (two average meals) a week of a variety of fish and shellfish that have lower mercury levels.

Shrimp, canned light tuna (not albacore tuna), salmon, pollock, and catfish are all regarded to have rather low concentrations of Hg. If advice for local consumption of contaminated fish is unavailable, it was recommended that people eat no more than two fish meals per week. Following a request from the European Commission, the European Food Safety Authority's (EFSA) Scientific Panel on Contaminants in the Food Chain (CONTAM) evaluated the possible risks to human health from consumption of foods contaminated with mercury, in particular MeHg; CONTAM used intake estimates for Europe. The Panel also considered the PTWI recently established by JECFA and the intake limits established by the U.S. National Research Council (US-NRC).

3.6 Risks and Benefits of Fish Consumption

It is well known that most human exposure to MeHg is through consumption of fish and shellfish. The levels of MeHg that reach human hair is dependent on the both the amount and the species of fish or other seafood consumed and the degree of their Hg contamination (Díez and Bayona 2002; Díez et al. 2007; Díez et al. 2008a, 2008b; Montuori et al. 2004, 2006). Several papers have recently dealt with the relative benefits and risks of fish consumption (Bouzan et al. 2005; Cohen et al. 2005a, b, c; Konig et al. 2005; McMichael and Butler 2005; Teutsch and Cohen 2005; Willett 2005). These articles address the quantity of fish people can consume relative to their corresponding risks and benefits. During late gestation, the developing brain is most vulnerable to neurochemical disruption from Hg exposure. Fetuses are known to face high-risks from MeHg exposure (Choi 1989; IPCS 1990) because of the susceptibility of the developing brain (IPCS 1990; Sakamoto et al. 2002); higher amounts of MeHg accumulate in cord blood than in maternal blood (Stern and Smith 2003; Vahter et al. 2000). Therefore, efforts must be made to protect fetuses from risks associated with MeHg exposure, particularly in populations such as the Japanese and other Asians, who consume large amounts of fish and other seafood (Agusa et al. 2005; Feng et al. 1998; Yasutake et al. 2004). Other populations that may face higher risks because they consume higher amounts of seafood are people in the Arctic region (Muckle et al. 2001;Van Oostdam et al.

1999), and others that dwell along rivers, lakes and coasts (Campbell et al. 2003; Dellinger 2004; Myers et al. 1997) such as Brazilians in the Tapajos River basin in the Amazon (Akagi et al. 1995; Lebel et al. 1997; Malm et al. 1995).

Fish constitute an important source of energy, protein and other nutrients (Clarkson and Strain 2003) and are low in saturated fats; they also contain essential nutrients such as heart healthy omega-3 fatty acids. Indeed, human intake of the n-3, longer chain polyunsaturated fatty acids (PUFAs), such as eicosapentaenoic acid and docosahexaenoic acid (DHA), is also known to occur mainly from fish consumption. Both of these fatty acids are very beneficial for human health (Skerrett and Hennekens 2003), particularly DHA, which is known to be important for normal brain function and development (Makrides et al. 1995; Yavin et al. 2001). A recent study of over 7,000 children revealed that when fish is not contaminated, moderate fish intake by the mother during pregnancy, and by the infant postnatally, may benefit brain development (Daniels et al. 2004). One study of neurodevelopment in infants suggested that maternal mercury exposure, and fish intake, had opposing effects on visually-mediated neurobehavioral tests (Oken et al. 2005). In a small Faroese birth cohort (Steuerwald et al. 2000), it was found that prenatal MeHg exposure adversely affected neonatal neurologic function, and, selenium and n-3 fatty acid status did not affect this outcome. Recently, Sakamoto et al. (2004) have reported a significant positive correlation between MeHg and DHA concentrations in fetal circulation. These last results confirm that both MeHg and DHA, which originated from fish consumption, transferred from maternal to fetal circulation to impart its positive effects on the fetus. This finding confirms that a decrease in fish consumption may cause decreases in MeHg and DHA levels. Pregnant women, in particular, should not give up eating fish at the risk of loosing such benefits. However, they would do well to consume smaller fish, which contain lower MeHg levels, thereby balancing the risks and benefits from fish consumption.

Fortunately, fish with a high content of beneficial fatty acids do not necessarily contain high mercury levels, and a prudent choice may therefore be possible (Budtz-Jørgensen et al. 2007). Nevertheless, an increase of omega-3 fatty acids in the diet has been promoted by the American Heart Association (AHA); AHA encourages consumption of omega-3 fatty acids from a variety of sources such as fatty fish (salmon) or plant sources. The AHA recommends a minimum of two servings of fatty fish per week to confer cardio protective effects (Krauss et al. 2001). Table 2 presents average Hg concentrations found in fish species popular with American consumers. As previously described, the health benefits of eating fish high in omega fatty acids are important for cardiovascular health and fetal development, in particular. Table 2 also provides the relative fatty acid content of fish popular with consumers.

3.7 Biomarkers and Exposure Evaluation

To determine the effect of MeHg on humans, it is preferred to use a biological indicator in the body that reflects the MeHg concentration in the major target organ, the brain (Cernichiari et al. 1995a). Blood and hair Hg concentrations are

Table 2 Relative fatty acid content and Hg concentration in popular fish

Species	Average Hg wt/wt (mg/kg)	Relative fatty acid content
Catfish	0.05	Low (channel) moderate (brown bullhead)
Clams	<0.01*	Low
Cod	0.095	Low
Crab (blue, king, snow)	0.06	Moderate
Flatfish (flounder, sole, plaice)	0.045	Low
Halibut	0.252	Moderate
Mackerel king	0.730	High
Pollock	0.041	Moderate
Salmon	0.014*	High
Scallop	0.05	Low
Shark	0.988	Low
Shrimp	<0.01*	Moderate
Swordfish	0.976	Low
Trout	0.072	Moderate (Rainbow) High (Lake)
Tuna, (canned, light)	0.118	Moderate
Tuna (canned, albacore)	0.353	High
Tuna (fresh/frozen, albacore)	0.357	Moderate
Tuna (fresh/frozen, bigeye)	0.639	Moderate
Tuna (fresh/frozen, skipjack)	0.205	Moderate
Tuna (fresh/frozen, yellowfin)	0.325	Moderate

Italics indicate freshwater fish. All other fish are marine species.
Mercury was measured as total mercury (THg), except for species (*), when only MeHg was analyzed
Source of data: FDA1990–2004, "National Marine Fisheries Service Survey of Trace Elements in the Fishery Resource" Report 1978, "The Occurrence of Mercury in the Fishery Resources of the Gulf of Mexico" Report 2000. http://www.cfsan.fda.gov/~frf/sea-mehg.html

used as valid biomarkers for MeHg in both the adult and fetal brain (in the latter case, cord blood or maternal hair), although each provides a somewhat different reflection of exposure (NRC 2000). Blood gives an estimate of exposure over the most recent 100–140 d, whereas, hair reflects the average exposure over the growth period of the segment. Whether hair or blood is a better indicator of fetal brain exposure has been debated for several yr. Some researchers argue that MeHg residues in cord blood is in closer contact with the fetal brain than is mercury in maternal hair; they also argue that hair is potentially subject to external contamination (Budtz-Jorgensen et al. 2004). Moreover, various types of hair treatments may reduce mercury levels in hair (Dakeishi et al. 2005). In contrast, proponents for using maternal hair argue that cord blood levels are only relevant at the time of delivery, whereas, hair recapitulates mercury levels throughout pregnancy. Moreover, because at least 80% of the MeHg blood is associated with the RBC (red blood cells), cord blood levels will be influenced by the hematocrit. Depending on the method used to collect cord blood, the hematocrit may vary widely (Clarkson and Magos 2006). With respect to hair treatment, a recent comprehensive study of mercury levels in human hair found no effects on hair mercury levels (McDowell et al. 2004). The mercury concentration in hair reflects the MeHg concentration in the blood during hair formation, and is frequently used as biomarker for evaluating MeHg exposure.

The mechanisms of MeHg transport have implications for the choice of indicator media. Because MeHg is highly mobile, and with continued exposures soon attains a steady state distribution in the body, levels of mercury in virtually any tissue or biological fluid may yield results useful in determining MeHg exposure. Biomarkers commonly used to assess MeHg exposure to the fetal brain include cord blood, placental tissue, and maternal blood and scalp hair.

Because of its proximity to the fetal brain, cord blood is generally the biomarker of first choice, however, based on what is known about the mechanisms of transport and disposition of MeHg, maternal scalp hair offers the best functional index of fetal brain levels. As mentioned, the MeHg-cysteine complex is responsible for MeHg transport into cells via the large neutral amino acid carrier. However, during its formation, the hair follicle has a high demand for amino acids as substrates for proteins, especially keratin. The large neutral amino-acid carriers will be highly active, not only in transporting its normal substrates, but also in transporting the MeHg–cysteine complex. Once transported into the hair follicular cells, keratin proteins synthesized in these cells have a high cysteine content that provides ample binding and stable storage for the transported MeHg. In fact, MeHg is incorporated into hair follicles in proportion to its content in blood. The hair-to-plasma ratio is about 2,500:1, whereas, the hair-to-blood ratio in humans is estimated to be ~250:1 expressed as mg/g hair to mg Hg/L blood (IPCS 1990). Once incorporated into hair, the mercury is stable, and can provide a history of exposure (Phelps et al. 1980; IPCS 1990). Hair grows at an approximate rate of about 1 cm/mon, and studies can be performed to determine past exposures if the length of the hair permits it. Because the half-life of MeHg in the body is about 1.5–2 mon (Smith and Farris 1996), the hair nearest the scalp best reflects current exposures and recent blood concentrations. A population that regularly and frequently consumes fish will show a clear correlation between total mercury (THg) content of hair and blood MeHg levels. Furthermore, MeHg levels in hair and THg are linearly related, with MeHg accounting for the 70–80% of hair THg (Cernichiari et al. 1995a; Dolbec et al. 2001). Finally, if the mechanisms of transport strongly suggest that the hair follicle accumulates the same transportable species of mercury as that which enters the brain, the levels of mercury in maternal hair, in a population of fish consumers, correlate highly with levels in the brain of newborn infants (Cernichiari et al. 1995a). The convenience of sampling and storing scalp hair is advantageous for monitoring and field studies and it has been extensively demonstrated that the THg concentration in hair reflects the average MeHg concentrations circulating in blood.

The growth rate of hair, generally estimated at 1 cm/mon, can have both inter- and intra-individual variability. Recent advances in analysis of a single hair strands (Duford et al. 2007; Legrand et al. 2004, 2007; Toribara 2001) should yield more information on the relationship between Hg uptake and Hg deposition in hair.

4 Summary

Mercury (Hg), and the organometallic compounds formed from it, are among the most toxic of substances to the global environment. Mercury is environmentally ubiquitous, and both wildlife and humans are exposed to the toxic effects of its environmental residues, primarily elemental mercury (Hg^0), divalent mercury (Hg^{2+}) and methylmercury (MeHg). Humans are exposed to different forms of Hg, and potential health risks have been reported from such exposures; examples of Hg exposure include mercury vapor from dental amalgams, occupational exposures and exposures during artisan and small-scale gold mining operations. Despite the significance of these foregoing Hg exposures, of particular concern is human and wildlife exposure to MeHg, a potent neurotoxicant. Once incorporated into the body, MeHg easily penetrates the blood-brain barrier and causes damage to the central nervous system, particularly in fetuses. It bioaccumulates and biomagnifies in the aquatic food chain; consequently, fish and seafood consumption is the major pathway by which humans are exposed to MeHg.

MeHg is the focus of this review. It adversely affects humans and is currently the subject of intense public health interest and worldwide concern. In this review, I summarize the sources and cycling of global mercury in the environment, pathways of exposure, toxicity and exposure evaluation, toxicokinetics, the common biomarkers to evaluate exposure and effects in populations, and finally review the nutritional risks and benefits from fish consumption.

References

Acha D, Iniguez V, Roulet M, Guimarães JRD, Luna R, Alanoca L, Sanchez S (2005) Sulfate-reducing bacteria in floating macrophyte rhizospheres from an Amazonian floodplain lake in Bolivia and their association with Hg methylation. *Appl Environ Microbiol* 71:7531–7535.

Adimado AA, Baah DA (2002) Mercury in human blood, urine, hair, and fish from the Ankobra and Tano River Basins in southwestern Ghana. *Bull Environ Contam Toxicol* 68:339–346.

Agusa T, Kunito T, Iwata H, Monirith I, Tana TS, Subramanian A, Tanabe S (2005) Mercury contamination in human hair and fish from Cambodia: Levels, specific accumulation and risk assessment. *Environ Pollut* 134:79–86.

Akagi H, Malm O, Branches FJP, Kinjo Y, Kashima Y, Guimaraes JRD, Oliveira RB, Haraguchi K, Pfeiffer WC, Takizawa Y, Kato H (1995) Human exposure to mercury due to goldmining in the Tapajos River Basin, Amazon, Brazil – Speciation of mercury in human hair, blood and urine. *Water Air Soil Pollut* 80:85–94.

Akagi H, Grandjean P, Takizawa Y, Weihe P (1998) Methylmercury dose estimation from umbilical cord concentrations in patients with Minamata disease. *Environ Res* 77:98–103.

Al-Majed NB, Preston MR (2000) Factors influencing the total mercury and methyl mercury in the hair of the fishermen of Kuwait. *Environ Pollut* 109:239–250.

ATSDR (1999) Toxicological Profile for Mercury. Agency for Toxic Substances and Disease Registry, Atlanta, GA.

Bakir F, Damluji SF, Amin-Zaki L, Murtadha M, Khalidi A, al-Rawi NY, Tikriti S, Dahahir HI, Clarkson TW, Smith JC, Doherty RA (1973) Methylmercury poisoning in Iraq. *Science* 181:230–241.

Ballatori N, Clarkson TW (1982) Developmental-changes in the biliary-excretion of methylmercury and glutathione. *Science* 216:61–63.

Barbosa AC, Jardim W, Dorea JG, Fosberg B, Souza J (2001) Hair mercury speciation as a function of gender, age, and body mass index in inhabitants of the Negro River Basin, Amazon, Brazil. *Arch Environ Contam Toxicol* 40:439–444.

Batista J, Schuhmacher M, Domingo JL, Corbella J (1996) Mercury in hair for a child population from Tarragona Province, Spain. *Sci Total Environ* 193:143–148.

Becker K, Kaus S, Krause C, Lepom P, Schulz C, Seiwert M, Seifert B (2002) German environmental survey 1998 (GerES III): Environmental pollutants in blood of the German population. *Int J Hyg Environ Health* 205:297–308.

Becker K, Schulz C, Kaus S, Seiwert M, Seifert B (2003) German environmental survey 1998 (GerES III): Environmental pollutants in the urine of the German population. *Int J Hyg Environ Health* 206:15–24.

Berg T, Sekkesaeter S, Steinnes E, Valdal AK, Wibetoe G(2003)Springtime depletion of mercury in the European Arctic as observed at Svalbard. *Sci Total Environ* 304:43–51.

Bergan T, Rodhe H (2001) Oxidation of elemental mercury in the atmosphere; Constraints imposed by global scale modelling. *J Atmos Chem* 40:191–212.

Bergan T, Gallardo L, Rodhe H (1999) Mercury in the global troposphere: A three-dimensional model study. *Atmos Environ* 33: 1575–1585.

Bernard S, Redwood L, Blaxill M (2004) Thimerosal, mercury, and autism: Case study in the failure of the risk assessment paradigm. *Neurotoxicol* 25:710–710.

Bjornberg KA, Vahter M, Petersson-Grawe K, Glynn A, Cnattingius S, Darnerud PO, Atuma S, Aune M, Becker W, Berglund M (2003) Methyl mercury and inorganic mercury in Swedish pregnant women and in cord blood: Influence of fish consumption. *Environ Health Perspect* 111:637–641.

Bouzan C, Cohen JT, Connor WE, Kris-Etherton PM, Gray GM, Konig A, Lawrence RS, Savitz DA, Teutsch SM (2005) A quantitative analysis of fish consumption and stroke risk. *Am J Prev Med* 29:347–352.

Budtz-Jørgensen E, Grandjean P, Jorgensen PJ, Weihe P, Keiding N (2004) Association between mercury concentrations in blood and hair in methylmercury-exposed subjects at different ages. *Environ Res* 95:385–393.

Budtz-Jørgensen E, Grandjean P, Weihe P (2007) Separation of risks and benefits of seafood intake. *Environ Health Perspect* 115:323–327.

Calasans CF, Malm O (1997) Elemental mercury contamination survey in a chlor-alkali plant by the use of transplanted Spanish moss, Tillandsia usneoides (L.). *Sci Total Environ* 208:165–177.

Campbell LM, Dixon DG, Hecky RE (2003) A review of mercury in Lake Victoria, East Africa: Implications for human and ecosystem health. *J Toxicol Environ Health-Part B-Crit Rev* 6:325–356.

Canuel R, de Grosbois SB, Atikesse L, Lucotte M, Arp P, Ritchie C, Mergler D, Chan HM, Amyot M, Anderson R (2006) New evidence on variations of human body burden of methylmercury from fish consumption. *Environ Health Perspect* 114:302–306.

Cernichiari E, Brewer R, Myers GJ, Marsh DO, Lapham LW, Cox C, Shamlaye CF, Berlin M, Davidson PW, Clarkson TW (1995a) Monitoring methylmercury during pregnancy: Maternal hair predicts fetal brain exposure. *Neurotoxicol* 16:705–709.

Cernichiari E, Toribara TY, Liang L, Marsh DO, Berlin MW, Myers GJ, Cox C, Shamlaye CF, Choisy O, Davidson PW, Clarkson TW (1995b) The biological monitoring of mercury in the Seychelles study. *Neurotoxicol* 16:613–627.

Clarkson TW (1993) Mercury—Major issues in environmental-health. *Environ Health Perspect* 100:31–38.

Clarkson TW, Magos L (2006) The toxicology of mercury and its chemical compounds. *Crit Rev Toxicol* 36:609–662.

Clarkson TW, Strain JJ (2003) Nutritional factors may modify the toxic action of methyl mercury in fish-eating populations. *J Nutr* 133:1539S–1543S.

Clarkson TW, Magos L, Myers GJ (2003) Human exposure to mercury: The three modern dilemmas. *J Trace Elem Exp Med* 16:321–343.

Crump KS, Kjellstrom T, Shipp AM, Silvers A, Stewart A (1998) Influence of prenatal mercury exposure upon scholastic and psychological test performance: Benchmark analysis of a New Zealand cohort. *Risk Anal* 18:701–713.

Choi BH (1989) The effects of methylmercury on the developing brain progress. *Neurobiol* 32:447–470.

Choi MH, Cech JJ (1998) Unexpectedly high mercury level in pelleted commercial fish feed. *Environ Toxicol Chem* 17:1979–1981.

Cohen JT, Bellinger DC, Connor WE, Kris-Etherton PM, Lawrence RS, Savitz DA, Shaywitz BA, Teutsch SM, Gray GM (2005a) A quantitative risk-benefit analysis of changes in population fish consumption. *Am J Prev Med* 29:325–334.

Cohen JT, Bellinger DC, Connor WE, Shaywitz BA (2005b) A quantitative analysis of prenatal intake of n-3 polyunsaturated fatty acids and cognitive development. Am J Prev Med 29:366–374.

Cohen JT, Bellinger DC, Shaywitz BA (2005c) A quantitative analysis of prenatal methyl mercury exposure and cognitive development. *Am J Prev Med* 29:353–365.

Cordier S, Garel M, Mandereau L, Morcel H, Doineau P, Gosme-Seguret S, Josse D, White R, Amiel-Tison C (2002) Neurodevelopmental investigations among methylmercury-exposed children in French Guiana. *Environ Res* 89:1–11.

Dakeishi N, Nakai K, Sakamoto M, Iwata T, Suzuki K, Liu X-J, Ohno T, Kurosawa T, Satoh H, Murata K (2005) Effects of hair treatment on hair mercury—The best biomarker of methylmercury exposure? *Environ Health Prev Med* 10:208–212.

Daniels JL, Longnecker MP, Rowland AS, Golding J (2004) Fish intake during pregnancy and early cognitive development of offspring. *Epidemiol* 15:394–402.

de Campos MS, Sarkis JES, Muller RCS, Brabo ED, Santos ED (2002) Correlation between mercury and selenium concentrations in Indian hair from Rondonia State, Amazon region, Brazil. *Sci Total Environ* 287:155–161.

Dellinger JA (2004) Exposure assessment and initial intervention regarding fish consumption of tribal members of the Upper Great Lakes Region in the United States. *Environ Res* 95:325–340.

Dickman MD, Leung CKM, Leong MKH (1998) Hong Kong male subfertility links to mercury in human hair and fish. *Sci Total Environ* 214:165–174.

Díez S, Bayona JM (2002) Determination of methylmercury in human hair by ethylation followed by headspace solid-phase microextraction-gas chromatography-cold-vapour atomic fluorescence spectrometry. *J Chromatogr A* 963:345–351.

Díez S, Montuori P, Querol X, Bayona JM (2007) Total mercury in the hair of children by combustion atomic absorption spectrometry (Comb-AAS). *J Anal Toxicol* 31:144–149.

Díez S, Montuori P, Pagano A, Sarnacchiaro P, Bayona JM, Triassi M (2008a) Hair mercury levels in an urban population from southern Italy: Fish consumption as a determinant of exposure. *Environ Int* 34:162–167.

Díez S, Delgado S, Aguilera I, Astray J, Pérez-Gómez B, Torrent M, Sunyer J, Bayona JM (2008b) Prenatal and early childhood exposure to mercury and methylmercury in Spain, a high-fish-consumer country. Arch Environ Contam Toxicol. In press. doi:10.1007/s00244–008–9213–7

Dolbec J, Mergler D, Passos CJS, de Morais SS, Lebel J (2000) Methylmercury exposure affects motor performance of a riverine population of the Tapajos river. *Braz Amazon Int Arch Occupat Environ Health* 73:195–203.

Dolbec J, Mergler D, Larribe F, Roulet M, Lebel J, Lucotte M (2001) Sequential analysis of hair mercury levels in relation to fish diet of an Amazonian population, Brazil. *Sci Total Environ* 271:87–97.

Dorea JG (2006) Fish meal in animal feed and human exposure to persistent bioaccumulative and toxic substances. *J Food Prot* 69:2777–2785.

Dorea JG, Barbosa AC, Ferrari I, De Souza JR (2003) Mercury in hair and in fish consumed by Riparian women of the Rio Negro, Amazon, Brazil. *Int J Environ Health Res* 13:239–248.

Dorea JG, Barbosa AC, Ferrari I, de Souza JR (2005) Fish consumption (hair mercury) and nutritional status of Amazonian Amer-Indian Children. *Am J Hum Biol* 17:507–514.

Drasch G, Bose-O'Reilly S, Beinhoff C, Roider G, Maydl S (2001) The Mt Diwata study on the Philippines 1999: Assessing mercury intoxication of the population by small scale gold mining. *Sci Total Environ* 267:151–168.

Duford DA, Lafleur JP, Lam R, Skinner CD, Salin ED (2007) Induction heating-electrothermal vaporization for direct mercury determination in a single human hair by atomic fluorescence and atomic absorption spectrometry. *J Anal At Spectrom* 22:326–329.

Easton MDL, Luszniak D, Von der Geest E (2002) Preliminary examination of contaminant loadings in farmed salmon, wild salmon and commercial salmon feed. *Chemosphere* 46:1053–1074.

Ebinghaus R, Kock HH, Temme C, Einax JW, Lowe AG, Richter A, Burrows JP, Schroeder WH (2002) Antarctic springtime depletion of atmospheric mercury. *Environ Sci Technol* 36:1238–1244.

Elhamri H, Idrissi L, Coquery M, Azemard S, El Abidi A, Benlemlih M, Saghi M, Cubadda F (2007) Hair mercury levels in relation to fish consumption in a community of the Moroccan Mediterranean coast. *Food Addit Contam* 24:1236–1246.

FDA (U.S. Food and Drug Administration) (1990–2004) Mercury levels in commercial fish and shellfish, at [http://www.cfsan.fda.gov/~frf/sea-mehg.html].

Feng QY, Suzuki Y, Hisashige A (1998) Hair mercury levels of residents in China, Indonesia, and Japan. *Arch Environ Health* 53:36–43.

Ferrari CP, Dommergue A, Boutron CF, Skov H, Goodsite M, Jensen B (2004) Nighttime production of elemental gaseous mercury in interstitial air of snow at Station Nord, Greenland. *Atmos Environ* 38:2727–2735.

Fitzgerald WF, Engstrom DR, Mason RP, Nater EA (1998) The case for atmospheric mercury contamination in remote areas. *Environ Sci Technol* 32:1–7.

Foran JA, Hites RA, Carpenter DO, Hamilton MC, Mathews-Amos A, Schwager SJ (2004)A survey of metals in tissues of farmed Atlantic and wild Pacific salmon. *Environ Toxicol Chem* 23:2108–2110.

Frery N, Maury-Brachet R, Maillot E, Deheeger M, de Merona B, Boudou A (2001) Gold-mining activities and mercury contamination of native Amerindian communities in French Guiana: Key role of fish in dietary uptake. *Environ Health Perspect* 109:449–456.

Gao Y, Yan CH, Tian Y, Wang Y, Xie HF, Zhou X, Yu XD, Yu XG, Tong SL, Zhou QX, Shen XM (2007) Prenatal exposure to mercury and neurobehavioral development of neonates in Zhoushan City, China. *Environ Res* 105:390–399.

Goering PL, Galloway WD, Clarkson TW, Lorscheider FL, Berlin M, Rowland AS (1992) Toxicity assessment of mercury-vapor from dental amalgams. *Fundam Appl Toxicol* 19:319–329.

Grandjean P, Weihe P, Jorgensen PJ, Clarkson TW, Cernichiari E, Videro T (1992) Impact of maternal seafood diet on fetal exposure to mercury, selenium, and lead. *Arch Environ Health* 47:185–195.

Grandjean P, Weihe P, White RF, Debes F, Araki S, Yokoyama K, Murata K, Sorensen N, Dahl R, Jorgensen PJ (1997) Cognitive deficit in 7-year-old children with prenatal exposure to methylmercury. *Neurotoxicol Teratol* 19:417–428.

Grandjean P, White RF, Nielsen A, Cleary D, Santos ECD (1999) Methylmercury neurotoxicity in Amazonian children downstream from gold mining. *Environ Health Perspect* 107:587–591.

Grandjean P, White RF, Weihe P (2004) Neurobehavioral epidemiology: Application in risk assessment. *Environ Health Perspect* 2:397–400.

Hac E, Krzyzanowski M, Krechniak J (2000) Total mercury in human renal cortex, liver, cerebellum and hair. *Sci Total Environ* 248:37–43.

Halsey NA (1999) Limiting infant exposure to thimerosal in vaccines and other sources of mercury. *J Am Med Assoc* 282:1763–1766.

Hansen G, Victor R, Engeldinger E, Schweitzer C (2004) Evaluation of the mercury exposure of dental amalgam patients by the Mercury Triple Test. *Occup Environ Med* 61:535–540.

Harada M (1995) Minamata disease—Methylmercury poisoning in Japan caused by environmental-pollution. *Crit Rev Toxicol* 25:1–24.

Harada M, Nakanishi J, Konuma S, Ohno K, Kimura T, Yamaguchi H, Tsuruta K, Kizaki T, Ookawara T, Ohno H (1998) The present mercury contents of scalp hair and clinical symptoms in inhabitants of the Minamata area. *Environ Res* 77:160–164.
Harada M, Nakachi S, Cheu T, Hamada H, Ono Y, Tsuda T, Yanagida K, Kizaki T, Ohno H (1999) Monitoring of mercury pollution in Tanzania: Relation between head hair mercury and health. *Sci Total Environ* 227:249–256.
Harada M, Nakachi S, Tasaka K, Sakashita S, Muta K, Yanagida K, Doi R, Kizaki T, Ohno H (2001) Wide use of skin-lightening soap may cause mercury poisoning in Kenya. *Sci Total Environ* 269:183–187.
Harakeh S, Sabra N, Kassak K, Doughan B (2002) Factors influencing total mercury levels among Lebanese dentists. *Sci Total Environ* 297:153–160.
Harnly M, Seidel S, Rojas P, Fornes R, Flessel P, Smith D, Kreutzer R, Goldman L (1997) Biological monitoring for mercury within a community with soil and fish contamination. *Environ Health Perspect* 105:424–429.
Hightower JM, Moore D (2003) Mercury levels in high-end consumers of fish. *Environ Health Perspect* 111:604–608.
Holsbeek L, Das HK, Joiris CR (1996) Mercury in human hair and relation to fish consumption in Bangladesh. *Sci Total Environ* 186:181–188.
IPCS (1990) Environmental Health Criteria 101. Methylmercury; World Health Organization, Geneva. http://www.inchem.org/documents/ehc/ehc/ehc101.htm.
Jaffe D, Prestbo E, Swartzendruber P, Weiss-Penzias P, Kato S, Takami A, Hatakeyama S, Kajii Y (2005) Export of atmospheric mercury from Asia. *Atmos Environ* 39:3029–3038.
JECFA (Joint FAO/WHO (2006) Expert Committee on Food Additives) http://www.chem.unep.ch/mercury/Report/JECFA-PTWI.htm.
Kajiwara Y, Yasutake A, Adachi T, Hirayama K (1996) Methylmercury transport across the placenta via neutral amino acid carrier. *Arch Toxicol* 70:310–314.
Kajiwara Y, Yasutake A, Hirayama K (1997) Strain difference in methylmercury transport across the placenta. *Bull Environ Contam Toxicol* 59:783–787.
Kehrig HA, Malm O, Akagi H (1997) Methylmercury in hair samples from different riverine groups, Amazon, Brazil. *Water Air Soil Pollut* 97:17–29.
Kerper LE, Ballatori N, Clarkson TW (1992) Methylmercury transport across the blood-brain-barrier by an amino-acid carrier. *Am J Physiol* 262:R761–R765.
King JK, Kostka JE, Frischer ME, Saunders FM (2000) Sulfate-reducing bacteria methylate mercury at variable rates in pure culture and in marine sediments. *Appl Environ Microbiol* 66:2430–2437.
Knobeloch L, Anderson HA, Imma P, Peters D, Smith A (2005) Fish consumption, advisory awareness, and hair mercury levels among women of childbearing age. *Environ Res* 97:220–227.
Konig A, Bouzan C, Cohen JT, Connor WE, Kris-Etherton PM, Gray GM, Lawrence RS, Savitz DA, Teutsch SM (2005) A quantitative analysis of fish consumption and coronary heart disease mortality. *Am J Prev Med* 29:335–346.
Kosatsky T, Przybysz R, Armstrong B (2000) Mercury exposure in Montrealers who eat St Lawrence River sportfish. *Environ Res* 84:36–43.
Krauss RM, Eckel RH, Howard B, Appel LJ, Daniels SR, Deckelbaum RJ, Erdman JW, Kris-Etherton P, Goldberg IJ, Kotchen TA, Lichtenstein AH, Mitch WE, Mullis R, Robinson KM, Wylie-Rosett J, St Jeor S, Suttie J, Tribble DL, Bazzarre TL (2001) Revision 2000: A statement for healthcare professionals from the nutrition committee of the American Heart Association. *J Nutr* 131:132–146.
Lamborg CH, Fitzgerald WF, O'Donnell J, Torgersen T (2002) A non-steady-state compartmental model of global-scale mercury biogeochemistry with interhemispheric atmospheric gradients. *Geochim Cosmochim Acta* 66:1105–1118.
Lebel J, Roulet M, Mergler D, Lucotte M, Larribe F (1997) Fish diet and mercury exposure in a riparian Amazonian population. *Water Air Soil Pollut* 97:31–44.
Lebel J, Mergler D, Branches F, Lucotte M, Amorim M, Larribe F, Dolbec J (1998) Neurotoxic effects of low-level methylmercury contamination in the Amazonian Basin. *Environ Res* 79:20–32.

Lee WC, Lee MJ, Lee SM, Kim JS, Bae CS, Park TK (2000) An observation on the mercury contents of scalp hair in the urban residents of South Korea. *Environ Toxicol Pharmacol* 8:275–278.

Legrand M, Lam R, Jensen-Fontaine M, Salin ED, Chan HM (2004) Direct detection of mercury in single human hair strands by laser ablation inductively coupled plasma mass spectrometry (LA-ICP-MS). *J Anal Atom Spectrom* 19:1287–1288.

Legrand M, Lam R, Passos CJS, Mergler D, Salin ED, Chan HM (2007) Analysis of mercury in sequential micrometer segments of single hair strands of fish-eaters. *Environ Sci Technol* 41:593–598.

Lindberg SE, Brooks S, Lin CJ, Scott KJ, Landis MS, Stevens RK, Goodsite M, Richter A (2002) Dynamic oxidation of gaseous mercury in the Arctic troposphere at polar sunrise. *Environ Sci Technol* 36:1245–1256.

Lindberg A, Bjornberg KA, Vahter M, Berglund M (2004) Exposure to methylmercury in non-fish-eating people in Sweden. *Environ Res* 96:28–33.

Mahaffey KR, Clickner RP, Bodurow CC (2004) Blood organic mercury and dietary mercury intake: National Health and Nutrition Examination Survey, 1999 and 2000. *Environ Health Perspect* 112:562–570.

Makrides M, Neuman M, Simmer K, Pater J, Gibson R (1995) Are long-chain polyunsaturated fatty acids essential nutrients in infancy? *Lancet* 345:1463–1468.

Malm O (1998) Gold mining as a source of mercury exposure in the Brazilian Amazon. *Environ Res* 77:73–78.

Malm O, Branches FJP, Akagi H, Castro MB, Pfeiffer WC, Harada M, Bastos WR, Kato H (1995) Mercury and methylmercury in fish and human hair from the Tapajos river basin, Brazil. *Sci Total Environ* 175:141–150.

Mason RP, Fitzgerald WF, Morel FMM (1994) The biogeochemical cycling of elemental mercury—Anthropogenic influences. *Geochim Cosmochim Acta* 58:3191–3198.

McDowell MA, Dillon CF, Osterloh J, Bolger PM, Pellizzari E, Fernando R, de Oca RM, Schober SE, Sinks T, Jones RL, Mahaffey KR (2004) Hair mercury levels in US children and women of childbearing age: Reference range data from NHANES 1999–2000. *Environ Health Perspect* 112:1165–1171.

McMichael AJ, Butler CD (2005) Fish, health, and sustainability. *Am J Prev Med* 29:322–323.

Mendola P, Selevan SG, Gutter S, Rice D (2002) Environmental factors associated with a spectrum of neurodevelopmental deficits. *Ment Retard Dev Disabil Res Rev* 8:188–197.

Montuori P, Jover E, Alzaga R, Díez S, Bayona JM (2004) Improvements in the methylmercury extraction from human hair by headspace solid-phase microextraction followed by gas-chromatography cold-vapour atomic fluorescence spectrometry. *J Chromatogr A* 1025:71–75.

Montuori P, Jover E, Díez S, Ribas-Fito N, Sunyer J, Triassi M, Bayona JM (2006) Mercury speciation in the hair of pre-school children living near a chlor-alkali plant. *Sci Total Environ* 369:51–58.

Mortada WI, Sobh MA, El-Defrawy MM, Farahat SE (2002) Reference intervals of cadmium, lead, and mercury in blood, urine, hair, and nails among residents in Mansoura city, Nile Delta, Egypt. *Environ Res* 90:104–110.

Morton J, Mason HJ, Ritchie KA, White M (2004) Comparison of hair, nails and urine for biological monitoring of low level inorganic mercury exposure in dental workers. *Biomarkers* 9:47–55.

Muckle G, Ayotte P, Dewailly E, Jacobson SW, Jacobson JL (2001) Prenatal exposure of the Northern Quebec iuit infants to environmental contaminants. *Environ Health Perspect* 109:1291–1299.

Murata K, Budtz-Jorgensen E, Grandjean P (2002) Benchmark dose calculations for methylmercury-associated delays on evoked potential latencies in two cohorts of children. *Risk Anal* 22:465–474.

Murata K, Sakamoto M, Nakai K, Weihe P, Dakeishi M, Iwata T, Liu X J, Ohno T, Kurosawa T, Kamiya K, Satoh H (2004) Effects of methylmercury on neurodevelopment in Japanese children in relation to the Madeiran study. *Int Arch Occup Environ Health* 77:571–579.

Myers GJ, Davidson PW, Shamlaye CF, Axtell CD, Cernichiari E, Choisy O, Choi A, Cox C, Clarkson TW (1997) Effects of prenatal methylmercury exposure from a high fish diet on developmental milestones in the Seychelles Child Development Study. *Neurotoxicol* 18:819–829.

Myers GJ, Davidson PW, Cox C, Shamlaye C, Cernichiari E, Clarkson TW (2000) Twenty-seven years studying the human neurotoxicity of methylmercury exposure. *Environ Res* 83:275–285.

Nakagawa, R (1995) Concentration of mercury in hair of Japanese people. *Chemosphere* 30:127–133.

NRC (2000) Toxicological effects of methylmercury. National Research Council, National Academy Press, Washington, DC.

Oken E, Wright RO, Kleinman KP, Bellinger D, Amarasiriwardena CJ, Hu H, Rich-Edwards JW, Gillman MW (2005) Maternal fish consumption, hair mercury, and infant cognition in a US cohort. *Environ Health Perspect* 113:1376–1380.

Olivero J, Johnson B, Arguello E (2002) Human exposure to mercury in San Jorge river basin, Colombia (South America). *Sci Total Environ* 289:41–47.

Phelps RW, Clarkson TW, Kershaw TG, Wheatley B (1980) Interrelationships of blood and hair mercury concentrations in a North-American population exposed to methylmercury. *Arch Environ Health* 35:161–168.

Pesch A, Wilhelm M, Rostek U, Schmitz N, Weishoff-Houben M, Ranft U, Idel H (2002) Mercury concentrations in urine, scalp hair, and saliva in children from Germany. *J Exp Anal Environ Epidemiol* 12:252–258.

Razagui IBA, Haswell SJ (2001) Mercury and selenium concentrations in maternal and neonatal scalp hair—Relationship to amalgam-based dental treatment received during pregnancy. *Biol Trace Elem Res* 81:1–19.

Renzoni A, Zino F, Franchi E (1998) Mercury levels along the food chain and risk for exposed populations. *Environ Res* 77:68–72.

Rowland I, Davies M, Grasso P (1977) Biosynthesis of methylmercury compounds by intestinal flora of rat. *Arch Environ Health* 32:24–28.

Saeki K, Fujimoto M, Kolinjim D, Tatsukawa R (1996) Mercury concentrations in hair from populations in Wau-Bulolo area, Papua New Guinea. *Arch Environ Contam Toxicol* 30:412–417.

Sager PR (2006) Mercury levels in Argentine newborns and infants after receipt of routine vaccines containing thimerosal. *Neurotoxicol Teratol* 28:426–426.

Sakamoto M, Kakita A, Wakabayashi K, Takahashi H, Nakano A, Akagi H (2002) Evaluation of changes in methylmercury accumulation in the developing rat brain and its effects: A study with consecutive and moderate dose exposure throughout gestation and lactation periods. *Brain Res* 949:51–59.

Sakamoto M, Kubota M, Liu XH, Murata K, Nakai K, Satoh H (2004) Maternal and fetal mercury and n-3 polyunsaturated fatty acids as a risk and benefit of fish consumption to fetus. *Environ Sci Technol* 38:3860–3863.

Santos ECO, Camara VM, Jesus IM, Brabo ES, Loureiro ECB, Mascarenhas AFS, Fayal KF, Sa GC, Sagica FES, Lima MO, Higuchi H, Silveira IM (2002a) A contribution to the establishment of reference values for total mercury levels in hair and fish in Amazonia. *Environ Res* 90:6–11.

Santos ECO, Jesus IM, Camara VM, Brabo E, Loureiro ECB, Mascarenhas A, Weirich J, Luiz RR, Cleary D (2002b) Mercury exposure in Munduruku Indians from the community of Sai Cinza, State of Para, Brazil. *Environ Res* 90:98–103.

Schroeder WH, Anlauf KG, Barrie LA, Lu JY, Steffen A, Schneeberger DR, Berg T (1998) Arctic springtime depletion of mercury. *Nature* 394:331–332.

Skerrett PJ, Hennekens CH (2003) Consumption of fish and fish oils and decreased risk of stroke. *Prev Cardiol* 6:38–41.

Smith JC, Farris FF (1996) Methyl mercury pharmacokinetics in man: A reevaluation. *Toxicol Appl Pharmacol* 137:245–252.

Stern AH, Gochfeld M, Weisel C, Burger J (2001) Mercury and methylmercury exposure in the New Jersey pregnant population. *Arch Environ Health* 56:4–10.
Stern AH, Smith AE (2003) An assessment of the cord blood: Maternal blood methylmercury ratio: Implications for risk assessment. *Environ Health Perspect* 111:1465–1470.
Steuerwald U, Weihe P, Jorgensen PJ, Bjerve K, Brock J, Heinzow B, Budtz-Jorgensen E, Grandjean P (2000) Maternal seafood diet, methylmercury exposure, and neonatal neurologic function. *J Pediatr* 136:599–605.
Teutsch SM, Cohen JT (2005) Health trade-offs from policies to alter fish consumption. *Am J Prev Med* 29:324–324.
Toribara TY (2001) Analysis of single hair by XRF discloses mercury intake. *Hum Exp Toxicol* 20:185–188.
Travnikov O (2005) Contribution of the intercontinental atmospheric transport to mercury pollution in the Northern Hemisphere. *Atmos Environ* 39:7541–7548.
Tsubaki T, Irukayama K (1977) Minamata Disease: Methylmercury Poisoning in Minamata and Niigata, Japan. New York: Elsevier.
USEPA (1997) Mercury Study Report to Congress. Office of Air Quality Planning and Standards and Office of Research and Development, EPA 452/R-97-0003, Washington, DC.
USEPA (2001) Water Quality Criterion for the Protection of Human Health: Methyl Mercury. EPA 0823-R-01-001. Washington, DC.
Vahter M, Akesson A, Lind B, Bjors U, Schutz A, Berglund M (2000) Longitudinal study of methylmercury and inorganic mercury in blood and urine of pregnant and lactating women, as well as in umbilical cord blood. *Environ Res* 84:186–194.
Van Oostdam J, Gilman A, Dewailly E, Usher P, Wheatley B, Kuhnlein H, Neve S, Walker J, Tracy B, Feeley M, Jerome V, Kwavnick B (1999) Human health implications of environmental contaminants in Arctic Canada: A review. *Sci Total Environ* 230:1–82.
Watras CJ, Morrison KA, Kent A, Price N, Regnell O, Eckley C, Hintelmann H, Hubacher T (2005) Sources of methylmercury to a wetland-dominated lake in northern Wisconsin. *Environ Sci Technol* 39:4747–4758.
Willett WC (2005) Fish—Balancing health risks and benefits. *Am J Prev Med:*320–321.
Wong CSC, Duzgoren-Aydin NS, Aydin A, Wong MH (2006) Sources and trends of environmental mercury emissions in Asia. *Sci Total Environ* 368:649–662.
Yasutake A, Matsumoto M, Yamaguchi M, Hachiya N (2003) Current hair mercury levels in Japanese: Survey in five districts. *Tohoku J Exp Med* 199:161–169.
Yasutake A, Matsumoto M, Yamaguchi M, Hachiya N (2004) Current hair mercury levels in Japanese for estimation of methylmercury exposure. *J Health Sci* 50:120–125.
Yavin E, Glozman S, Green P (2001) Docosahexaenoic acid sources for the developing brain during intrauterine life. *Nutr Health* 15:219–224.

Waterborne Adenovirus

Kristina D. Mena (✉) and Charles P. Gerba

Contents

1 Introduction ... 134
 1.1 Taxonomy of Human Adenoviruses ... 134
 1.2 Structure and Physical/Chemical Properties ... 135
2 Human Diseases Associated with Adenoviruses ... 135
 2.1 Gastroenteritis ... 138
 2.2 Respiratory Infections ... 139
 2.3 Acute Respiratory Disease of Military Recruits ... 139
 2.4 Adenoviral Pneumonias ... 140
 2.5 Pharyngoconjunctival Fever ... 140
 2.6 Eye Infections ... 141
 2.7 Obesity ... 141
 2.8 Acute Hemorrhagic Cystitis ... 142
 2.9 Meningoencephalitis ... 142
 2.10 Other Diseases ... 142
3 Morbidity and Mortality ... 142
 3.1 Impact on Children ... 143
 3.2 Impact on the Immunocompromised ... 146
4 Waterborne Disease Outbreaks Associated with Adenoviruses ... 147
 4.1 Recreational Outbreaks ... 147
 4.2 Drinking Water Outbreaks ... 148
5 Occurrence and Survival of Adenoviruses in Water ... 148
 5.1 Occurrence in Sewage ... 149
 5.2 Occurrence in Surface Waters ... 150
 5.3 Occurrence in Groundwater ... 150
 5.4 Occurrence in Drinking Water ... 151
 5.5 Survival in the Environment ... 151
6 Economic Impact of Adenovirus Infections ... 152
7 Risk Assessment ... 152
 7.1 Dose-Response ... 152
 7.2 Previous Risk Assessments ... 152
 7.3 Risks for Drinking Water ... 154

K.D. Mena
University of Texas, Houston School of Public Health, Houston, Texas

C.P. Gerba
University of Arizona, Tucson, Arizona

D.M. Whitacre (ed.) *Reviews of Environmental Contamination and Toxicology*, Vol 198
doi: 10.1007/978-0-387-09646-9,

8 Removal of Adenoviruses by Water Treatment 155
9 Disinfection 155
10 Data Gaps 157
11 Summary 158
References 159

1 Introduction

The Environmental Protection Agency (EPA) is required, under provisions of the Safe Drinking Water Act (amended in 1996), to publish a list of unregulated contaminants, known or expected to occur in public water systems, that may pose a risk in drinking water (National Research Council 1999). In 1998, the first of these lists was produced, and is referred to as the Drinking Water Contaminant Candidate List, or CCL. This Drinking Water CCL included ten microbial contaminants, including the adenoviruses. The objective of this document is to present a review of the literature, in which we assess the health risks and economic burden associated with adenoviruses in drinking water.

1.1 Taxonomy of Human Adenoviruses

Rowe et al. (1953) first recognized adenoviruses as they searched for the cause of the common cold. The viruses were discovered in degenerating primary tissue culture cells originating from human adenoids and tonsils. One yr later, Hilleman and Werner (1954) identified similar viral agents in secretions from army recruits with acute respiratory illnesses. As the viruses were isolated, they were named according to their disease presentation. These included adenoid degeneration, adenoid-pharyngeal conjunctival and acute respiratory disease (ARD) viruses (Horwitz 2001). In 1956, these agents were recognized as being related entities and the name adenovirus was adopted, denoting the tissue in which these viruses were first discovered (Enders et al. 1956).

Human adenoviruses belong to the family *Adenoviridae*. The family classification is subdivided into the *Mastadenovirus* and *Aviadenovirus* genera (van Regenmortel et al. 2000). The *Mastadenovirus* genus includes all of the species infecting humans as well as the simian, murine, bovine, equine, porcine, ovine and canine species, and those infecting opossums. The *Aviadenovirus* genus only includes viruses that infect avian species. In general, the natural host ranges of adenoviruses are confined to one species or closely related species. For example, human adenoviruses fail to productively infect cells of monkey origin unless a co-infection with SV40 (papovavirus) is present (Fenger 1991).

Currently, there are 51 identifiable human adenovirus serotypes (Ad1–Ad51). These are divided into six subgenera (A–F) and four hemagglutination groups

Table 1 Human adenovirus serotype classification[a]

Subgroup	Serotype	Hemagglutination group
A	12, 18, 31	IV (Little or no agglutination)
B	3, 7, 11, 14, 16, 21, 34, 35, 50	I (Complete agglutination of Rhesus monkey erythrocytes)
C	1, 2, 5, 6	III (Partial agglutination of rat erythrocytes)
D	8–10, 13, 15, 17, 19, 20, 22–30, 32, 33, 36–39, 42–49, 51	II (Complete agglutination of rat erythrocytes)
E	4	III
F	40 and 41	III

[a]Adapted from Shenk (2001), van Regenmortel et al. (2000)

(I–IV). Each serotype is distinguished by its resistance to neutralization by antisera to other known adenovirus serotypes (Shenk 2001). Serotypes, designated on their oncogenicity and hemagglutination groups, are based on their ability to agglutinate rhesus monkey and rat erythrocytes (Foy 1997). Table 1 outlines the current classification scheme for human adenovirus serotypes.

1.2 Structure and Physical/Chemical Properties

Adenoviruses have a non-enveloped, icosahedral virion that consists of a core containing linear double-stranded DNA (26–45 kbp) enclosed by a capsid (Enriquez 2002). The capsid is composed of 252 capsomers, 240 of which are hexons and 12 of which are pentons. Each penton projects a single fiber that varies in length for each serotype, an exception being the enteric adenoviruse (EAd) pentons (serotypes 40 and 41) that project two fibers (Shenk 2001). Adenoviruses are ~70–100 nm in diameter.

As a result of their physical, chemical and structural properties, adenoviruses may survive extended periods of time outside host cells. They are stable in the presence of many physical and chemical agents, as well as adverse pH conditions. For example, adenoviruses are resistant to lipid solvents because they lack lipids within their structure (Liu 1991). Infectivity is optimal between pH 6.5 and 7.4; however, the viruses can withstand pH ranges between 5.0 and 9.0. In addition, adenoviruses are heat-resistant (particularly Ad4) and may remain infectious after freezing (Foy 1997).

2 Human Diseases Associated with Adenoviruses

Adenovirus-associated illnesses may have been documented as early as 1926 (Enriquez 2002). Routes of infection include the mouth, nasopharynx, and the ocular conjunctiva. Less frequently, the virus may become systemic and affect the bladder,

Table 2 Common illnesses associated with human adenoviruses[a]

Disease	Individuals at risk	Principal serotypes
Acute febrile pharyngitis	Infants, young children	1–3, 5–7
Pharyngoconjunctival fever	School-aged children	3, 7, 14
Acute respiratory disease	Military recruits	3, 4, 7, 14, 16, 21
Pneumonia	Infants, young children Military recruits	1–3, 4, 6, 7, 14, 16
Epidemic keratoconjunctivitis	Any	8–11,13, 15, 17, 19, 20, 22–29, 37
Follicular conjunctivitis	Infants, young children	3, 7
Gastroenteritis/Diarrhea	Infants, young children	18, 31, 40, 41
Urinary tract Colon Hepatitis	Bone marrow, liver or kidney transplant recipients, AIDS victims or immunosuppressed	34, 35 42–49 1, 2, 5

AIDS acquired immune deficiency syndrome
[a]Adapted from Enriquez (2002), Horwitz (2001)

liver, pancreas, myocardium or central nervous system (Horwitz 2001). Of the 51 known human serotypes, only a third are associated with human disease (Table 2). Other infections remain asymptomatic.

Adenoviruses are associated with a variety of clinical illnesses involving almost every human organ system. Illnesses induced by adenoviruses include upper (pharyngitis and tonsillitis) and lower (bronchitis, bronchiolitis and pneumonia) respiratory illnesses, conjunctivitis, cystitis and gastroenteritis. Several studies have reported that the EAds are second only to rotaviruses as the causative agents of acute gastroenteritis in infants and young children (Bates et al. 1993; Scott-Taylor and Hammond 1995; Shinozaki et al. 1991; Uhnoo et al. 1984; Wadell 1994). Fig. 1 illustrates human health outcomes associated with adenovirus infections.

Most illnesses caused by adenoviruses are acute and self-limiting. Although the symptomatic phase may be short, all adenoviruses can remain in the gastrointestinal tract and continue to be excreted for an extended period of time. Species within subgenera C may continue to be excreted for mon or even yr after disease symptoms have resolved. Adenoviruses can remain latent in the body (in tonsils, lymphocytes and adenoidal tissues) for yr and be reactivated under certain conditions, such as when there is a change in immune status. The long-term effect of such latent infections is unknown (Foy 1997).

Adenovirus infections may be accompanied by diarrhea, though the virus may be excreted even if diarrhea is not present (Wadell 1994). A large proportion of infections caused by subgenera A and D tend to be asymptomatic, whereas the species within subgenera B and E tend to result in a higher rate of symptomatic respiratory illnesses. Immunity is species-specific. The presence of pre-existing antibodies resulting from a previous infection is usually protective and, in such cases, symptomatic infection is rare (Foy 1997). Several seroepidemiology studies have been conducted to determine the prevalence of different serotypes of adenovirus. It has been estimated that, in the United States (US), 40–60% of children have antibodies to Ad1, Ad2 and Ad5 (Brandt et al. 1969). Within this

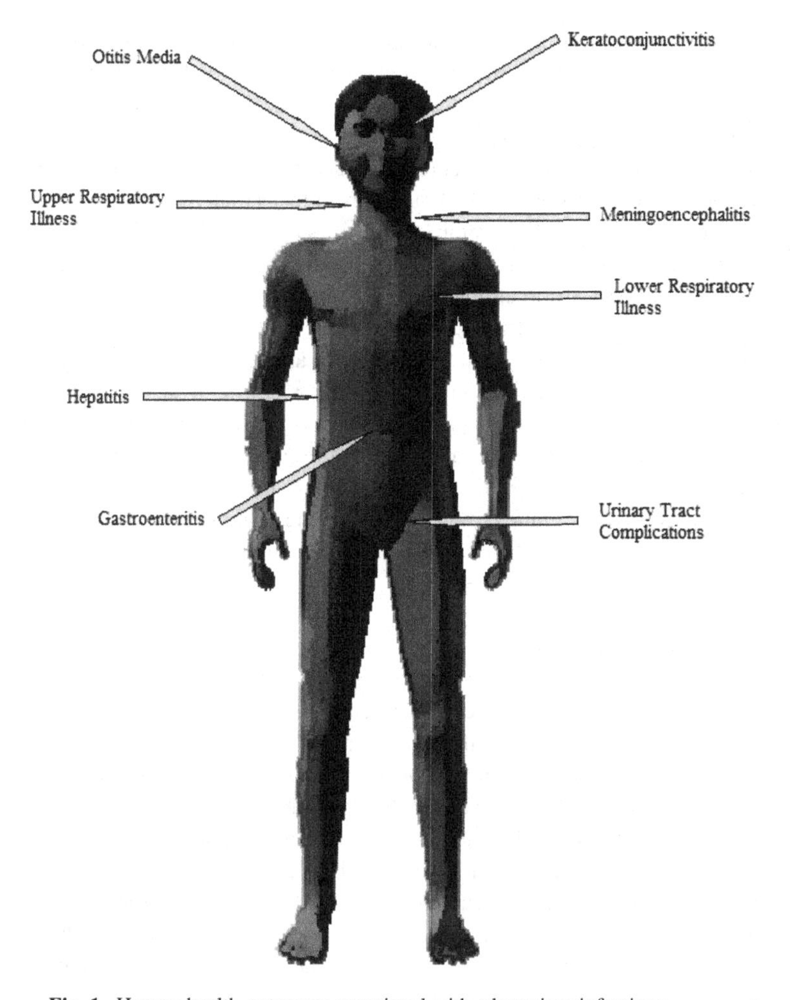

Fig. 1 Human health outcomes associated with adenovirus infections

population, there is a low incidence of antibodies to Ad3, Ad4 and Ad7, resulting in adults that are more susceptible to these serotypes (Singh-Naz and Rodriguez 1996). More infections with Ad3, Ad4 and Ad7 occur in adults than with Ad1, Ad2 and Ad5. This implies that there is long-lasting immunity to Ad1, Ad2 and Ad5. In studies of children, Ad1, Ad2, Ad5 and Ad6 (subgenera C) appear to be endemic, whereas other species (especially Ad3 and Ad7) tend to be epidemic or sporadic in nature.

It is difficult to confidently link adenoviruses to specific illnesses because asymptomatic, healthy people can shed viruses (Foy 1997). Occurrence studies comparing infection in healthy and ill people have found between 0% and 20% of asymptomatic people can shed adenovirus. The two primary viral surveillance studies to date were conducted in New York and Seattle, and provided information on the epidemiology of adenoviruses in these and similar communities (Cooney et al. 1972; Fox et al. 1969, 1977). In both cities, the respective study population was comprised of families with children. Respiratory and fecal specimens were examined and adenoviruses (Ad1, Ad2, Ad3, Ad5 and Ad7) were recovered more frequently than the other viruses studied (enteroviruses, rhinovirus and herpes virus). Adenoviruses were more commonly isolated from feces than respiratory secretions. In the Seattle study, an overall morbidity ratio of 0.49 was observed (Fox et al. 1977). The investigators also concluded that adenoviruses caused 5% of all infectious illnesses in infants and 3% in children aged 2– 4 yr.

Certain species of adenoviruses are oncogenic (from subgenera A and B), yet only when a human virus is introduced into an animal model (Foy 1997). Investigations into possible human cancer effects, through searches for tumor antigens and DNA sequences, have been negative (Mackey et al. 1979; Wold et al. 1979), though this research is outdated. Durepaire et al. (1995) suggest that these species may, at least theoretically, be associated with the development of tumors in patients with acquired immune deficiency syndrome (AIDS).

2.1 Gastroenteritis

The incidence of adenovirus induced-gastroenteritis in the world has ranged from 1.55 to 12%. EAds are second only to rotaviruses as leading causes of childhood gastroenteritis (Shinozaki et al.1991; Wadell 1994). Diarrhea is usually associated with fever and can last for up to 2 week. Although diarrhea can occur during infection by any type of adenovirus, Ad40 and Ad41 of subgenus F specifically cause gastroenteritis and diarrhea. Ad31 is also suspected of causing infantile gastroenteritis (Adrian et al. 1987) and, after detailed amino-acid sequence characterization, has been determined to be closely related to Ad40 and Ad41 (Pring-Åkerblom and Adrian 1995). Ad31 has been more closely associated with diarrhea than any of the other non-enteric adenoviruses (Foy 1997; LeBaron et al. 1990; Turner et al. 1987). A Canadian study of stool samples from ill children, conducted between 1983 and 1986, found 18% of adenovirus infections were caused by Ad31, 16.9% by Ad40 and 38% by Ad41 (Brown 1990). Similarly, a retrospective study in Toronto found that Ad31 represented 17% of 105 identified adenovirus cases (Krajden et al. 1990). The clinical syndrome caused by Ad31 is indistinguishable from that of the EAds.

Respiratory symptoms can sometimes occur with Ad40 and Ad41, but not often (Wadell 1994). Some estimate that Ad40/41 contributes from 5% to 20% of hospitalizations for diarrhea in developed countries (LeBaron et al.1990). Children, younger than a few yr, are the most vulnerable to infection (LeBaron et al. 1990; Lew et al. 1991). Some reports have shown the highest occurrence of infection in children exists for those less than 6 mon old (Bates et al. 1993; de Jong et al. 1993).

A prospective study of children enrolled in d-care centers in Texas generated data elucidating the role of EAds in group settings (Van et al. 1992). Children beginning at 6–24 mon old were monitored over a 5-yr period. Ten outbreaks, affecting 249 children, were associated with EAds. The infection rate during these ten outbreaks ranged from 20% to 60% (mean of 38%); however, 46% of the infected children remained asymptomatic. In another d-care study, 565 samples were taken from children with diarrhea and 129 samples from healthy children, primarily under the age of 35 mon (Lew et al. 1991). Of these samples, adenoviruses (any serotype) were identified in 8% of both well and ill children; EAds were identified in 2% of both groups. This indicated that ~50% of the infected children developed illness.

2.2 *Respiratory Infections*

Adenoviruses, particularly Ad1 through Ad6, have been shown to contribute to the etiology of acute respiratory diseases throughout the world, especially in children (Mizuta et al. 1994; Murtagh et al. 1993; Ray et al. 1993; Schmitz et al. 1983). Over 5% of respiratory illnesses in children younger than 5 yr of age are a result of adenovirus infections, and serologic surveys estimate that 10% of all respiratory diseases in children are caused by adenoviruses (Brandt et al. 1969). Symptoms such as fever, chills, headache, malaise and myalgia are commonly observed. Adenovirus transmission is initially through the nasopharynx. Adenoviruses can be recovered from the throat or stool of an infected child for up to 3 week (Horwitz 2001), and secondary transmission in households can be as high as 50% as a result of fecal-oral transmission from children shedding virus in the feces. As mentioned previously, respiratory symptoms can also occur with Ad40 and Ad41 infection, although this is not common (Wadell 1994). In addition, adenovirus respiratory infections are well documented in adults.

Ad1 through Ad7 are associated with upper respiratory illnesses such as pharyngitis, tonsillitis, and the "common cold." Lower respiratory illnesses from adenoviruses include bronchitis, bronchiolitis and pneumonia, with the latter sometimes resulting in severe (sometimes fatal) illness in infants and children (Abzug and Levin 1991). In a recent epidemiological study of pneumonia in hospitalized adults, 4% of 51 cases were caused by adenoviruses (Freymuth et al. 2004).

In children, adenovirus infections usually result in pharyngitis or tracheitis, but Ad7 has been reported to cause pneumonia and fulminant bronchiolitis (Baum 2000). Ad7 has also been implicated in severe infections resulting in mortality (Murtagh et al. 1993). In one study, 10 of 29 cases in children less than 5 yr of age infected with Ad7, died from subsequent pneumonia and necrotizing bronchiolitis (Murtagh et al. 1993).

2.3 *Acute Respiratory Disease of Military Recruits*

ARD is a respiratory infection characterized by nasal congestion, coryza (nasal discharge), and a cough found simultaneously with fever, chills, malaise, myalgia and a headache (Horwitz 2001). The disease may progress and become fatal due to pneumonitis. It can be transmitted from person-to-person, usually under conditions of fatigue and crowding, and was thus initially recognized in military recruits. Because of the disruption caused by this disease, a commission was organized to define this illness during World War II (Foy 1997). Through epidemiologic and human volunteer investigations, the disease was termed "ARD" for acute respiratory disease and was distinguished from other respiratory tract diseases as being caused by a filterable agent and as having an incubation period of 5–6 d (Dingle and Langmuir 1968).

Vaccines against Ad4 and Ad7 became available in 1971 and drastically reduced ARD in military institutions. In 1996, the manufacturer of these vaccines ceased production, resulting in an ARD epidemic, caused by Ad4, in more than 1,000 military trainees (McNeill et al. 1999). Deaths from adenovirus pneumonia in the military are rare, but do occur (Centers for Disease Control and Prevention, CDC 2001). Studies have shown that other risk settings include college dorms (CDC 1998), children's institutions, and d-care facilities.

2.4 *Adenoviral Pneumonias*

Although ARD has the potential to progress to pneumonitis, the incidence is rare. Ad3 and Ad7 have been responsible for outbreaks of severe or fatal pneumonia in infants and young children, as well as in military recruits (Dudding et al. 1972; Farng et al. 2002). Symptoms of adenoviral pneumonia include fever, cough, dyspnea and wheezing (Liu 1991).

2.5 *Pharyngoconjunctival Fever*

Pharyngoconjunctival fever (PCF) refers to a syndrome of pharyngitis, conjunctivitis, and spiking fever (Foy 1997). Symptoms of this syndrome include unilateral or bilateral conjunctivitis, mild throat tenderness, and fevers up to 104°F. The illness usually lasts from 5 to 7 d, but does not produce permanent eye damage (Liu 1991). Some cases may progress and result in pneumonia. The most commonly isolated adenovirus serotype is Ad3, although Ad7 and Ad14 have also been isolated (Horwitz 2001). The disease is best known for its association with summer camps, pools, and small lakes (D'Angelo et al. 1979; Harley et al. 2001). Transmission of the agent appears to require direct contact with the water, which contact allows the virus access to the eyes or upper respiratory tract. Secondary spread is common, although adults contracting the disease tend to have milder symptoms (usually only conjunctivitis).

2.6 Eye Infections

Infections of the conjunctiva refer to a clear membrane that coats the inner aspect of the eyelids and the outer surface of the eye. Conjunctivitis can occur sporadically or in large groups. Often, follicular conjunctivitis is contracted by swimming in inadequately chlorinated swimming pools, or in lakes during the summer (Horwitz 2001). The first documented outbreak of conjunctivitis was in 1955, when adenovirus was isolated from swimmers at a local pool; many outbreaks have been documented since (Martone et al. 1980; Papapetropoulou and Vantarakis 1998). Most cases result in only mild illness and complete recovery. Ad3 and Ad7 are the most commonly isolated species, although many other serotypes have been associated with this syndrome (Horwitz 2001).

Epidemic keratoconjunctivitis or EKC is a syndrome that causes inflammation of the conjunctiva and cornea. EKC was once referred to as "Shipyard Eye," because it was first described in shipyard workers (Horwitz 2001). Transmission was probably through medical facilities that treated eye trauma sustained on the job. EKC is considered highly contagious and begins with edema of the eyelids, pain, shedding tears, and photophobia. Some syndromes may progress to hemorrhagic conjunctivitis. The predominant adenovirus serotype currently associated with outbreaks of EKC is Ad37 (Kemp et al. 1983), although Ad8 and Ad19 have also been isolated from infected tissue. EKC outbreaks are commonly reported from offices of ophthalmologists (D'Angelo et al. 1981; Keenlyside et al. 1983; Kemp et al. 1983; Koo et al. 1989). The spread of EKC is thought to occur through insufficient sterilization of equipment or contact lenses (Kowalski et al. 2001), direct manipulation of the eye, or by use of eye solutions and ointments. Secondary spread between children and family members has also been documented, although direct inoculation into the eye appears to be necessary for disease (McMinn et al. 1991). In addition, sexual transmission of the virus may occur because of sporadic cases of EKC in young adults (Foy 1997).

2.7 Obesity

There is accumulating evidence that several viruses may be involved in animal and human obesity (Jaworowska and Barylak 2006). Studies in chickens, mice, and nonhuman primates indicate that Ad36 can cause obesity (Greenway 2006). Obese humans have a higher prevalence of serum antibodies to Ad36 than do lean humans (Atkinson et al. 2005). Other adenoviruses are capable of causing obesity in animals, but no correlation with antibodies has been demonstrated (Greenway 2006). The metabolic and molecular mechanisms of how adenovirus infections cause obesity are not precisely understood; however, increases in food intake alone cannot explain the observed increases in adiposity (tendency to store fat), suggesting that Ad36 induces metabolic changes (Vangipuram et al. 2004). One mechanism appears to be that Ad36 influences the differentiation of preadipocyte cells and/or the accumulation of lipids by adipocytes (cells that accumulate fat in the body) (Vangipuram et al. 2004).

2.8 *Acute Hemorrhagic Cystitis*

Ad11 and Ad21 are associated with acute hemorrhagic cystitis in patients following renal transplantation (Blohme et al. 1992). According to Hierholzer (1992), over 11% of transplant recipients become infected with adenoviruses, with an 18% case-fatality rate. The virus may be contracted as a result of the subsequent immunosuppression therapy, or introduced with the transplanted kidney. The cystitis is self-limiting and adenovirus can be isolated in the urine of 70% of immunocompromised patients with the disease (Numazaki et al. 1973). This illness most often occurs in boys and is associated with gross hematuria, or blood in the urine (Horwitz 2001).

2.9 *Meningoencephalitis*

The isolation of adenoviruses from the central nervous system of healthy individuals is rare. Cases involving virus in the cerebrospinal fluid or the brain are most often a result of host immunosuppression. The predominant adenovirus serotypes are Ad3, Ad5, Ad6, Ad7 and Ad12. One case of Ad32 isolated from the brain of a patient with malignant lymphoma has also been documented (Horwitz 2001).

2.10 *Other Diseases*

Adenoviruses have, at one time or another, been implicated as a cause of pericarditis (Rahal et al. 1976), chronic interstitial fibrosis (Kawai et al. 1976), rubelliform illness (Gutekunst and Heggie 1961) and congenital anomalies (Evans and Brown 1963). Although adenoviruses may be involved in these syndromes, their significance remains unclear.

3 Morbidity and Mortality

Because adenovirus is not a reportable disease agent, there are no national or population-based morbidity and mortality data available; most of the epidemiological data come from the study of select populations who appear to be most affected by adenovirus exposure. These include children in institutions such as

Table 3 Mortality ratios associated with adenovirus illnesses in sub-populations

Sub-population	Mortality ratio (%)	Reference
Immunocompromised patients	50–60	Shields et al. (1985); Zahradnik et al. (1980)
Immunocompromised patients	48	Hierholzer (1992)
Children <5 yr	34	Murtagh et al. (1993)
Immunocompromised children	83	Munoz et al. (1998)
Bone marrow transplant patients	1	Baldwin et al. (2000)
Immunocompromised patients	75	Pham et al. (2003)
Bone marrow transplant patients (children)	19	Kampmann et al. (2005)
Bone marrow transplant patients	>25	Krilov (2005)

hospitals and d-care centers, military recruits, immunocompromised individuals, and family groups. Adenovirus infections were common among military personnel in the 1950s and 1960s, with infection rates as high as 10%. They were also responsible for 90% of pneumonia hospitalizations (Buescher 1967). The impact of adenoviruses subsided with the implementation of a vaccine; however, the sole vaccine manufacturer ceased its production in 1996. Subsequently, morbidity ratios have again increased to 10–12%, and deaths have been observed in previously healthy military recruits (CDC 2001). In the following sections the various morbidity ratios observed during adenovirus outbreaks among children and the immunocompromised are discussed. Table 3 provides documented mortality ratios for select sub-populations.

3.1 Impact on Children

Numerous studies have been conducted to evaluate the occurrence of adenovirus infections in children (Table 4). Enteric infection in children results in disease 50% of the time. This percentage is greater when the infection is centered in the respiratory tract (Foy 1997). Attack rates for waterborne outbreaks have been as high as 67% in children, with secondary attack rates (person-to-person transmission) of 19% for adults and 63% for children (Foy et al. 1968). In an outbreak of adenovirus gastroenteritis among young children, an attack rate as high as 70% was observed (Richmond et al. 1979). As shown in Table 4, the peak incidence of most enteric adenoviral illnesses is in children less than 2 yr of age, although all age groups are affected.

Researchers, in a case-control study at an outpatient clinic in Baltimore, investigated the roles of potential causative agents of acute diarrhea in children less than 2 yr of age. Of the 246 cases, 26% were a result of viral pathogens. Only adenovirus and rotavirus were significantly associated with diarrhea when the control group was considered (Kotloff et al. 1988). Cases were more likely to be subject to household crowding, low birth weight and low-level maternal education (Kotloff et al. 1988).

Table 4 Selected studies on the occurrence of adenovirus in children[a]

Place	No.	% Total Adenovirus	% Enteric Adenovirus (Ad40/Ad41)	Other	Study
Sweden	200 well 416 ill	1.5 13.5	0 7.9	In- and out-patient	Uhnoo et al. 1984
Washington	270: 134 children; 136 adults	6.7	1.1	EAd (enteric adenovirus) cases all <24 mon with gastroenteritis	Rodriguez et al. (1985)
Baltimore	372 well 538 ill	–	1.3 5.2	2-yr prospective study; children <2 yr	Kotloff et al. (1988)
Guatemala	191 well 385 ill	–	4.7 14.0 of total well/ill samples	Ages 0–3 yr; of 59 hospit-alized: 51% rotavirus; 31.2% EAd	Cruz et al. (1990)
Korea	90 well 345 ill	–	2 9	6% with rota-virus and Ad 40/41 combined; 94% with EAd <24 mon	Kim et al. (1990)
Arizona	129 well 565 ill	8 8	2 2	4% ill, 1% well = astro-virus; 100% of Ad infected children <35 mon	Lew et al. (1991)
Finland	248 ill	4	–	26% rotavirus; 4% bacterial; 2/3 unidentified; sickest children-75% rotavirus, 0–2.5 yr	Ruuska and Vesikari (1991)
Sweden	50 well adults 50 ill children 50 ill adults	18 50 24	0 0 0	PCR methodology with 40/41 primers; mean age 32 yr, 21 mon, and 31 yr, respectively	Allard et al. (1991)
Argentina	766 well 180 ill	14.4 13.3	0.8 33.0 of total Ad	Family-based study of children <15 yr	Mistchenko et al. (1992)

(continued)

Table 4 (continued)

Place	No.	% Total Adenovirus	% Enteric Adenovirus (Ad40/Ad41)	Other	Study
England	1426 ill	17.8	16.4 of total Ad	78.3% rotavirus; 7.9% astro-virus; hospita-lized children <5 yr	Bates et al. (1993)
Rome	417 ill	7	–	18.2% rotavirus, astrovirus, hospitalized children	Donelli et al. (1993)
Brazil	79 well 67 ill	11.4 10	– 43.0 of total Ad	Hospitalized children <2 yr	Harsi et al. (1995)
Australia	4473 ill	–	3.2 (40/41) = 14.0 (40) and 86.0 (41) of total Ad	Children hospitalized with acute gastroenteritis	Grimwood et al. (1995)
England	452 ill	32 (non 40/41)	22.0 (40); 46.0 (41) total Ad	50% of positive samples were from infants <1 yr	Bryden et al. (1997)
China	44 ill	100	58.0 (40); 32.0 (41); 16.0 (both)	100 Ad infected population, EAds predominated in children <3 yr	Wang and Chen (1997)

PCR polymerase chain reaction, *Ad* adenovirus, *EAd* enteric adenovirus
[a]Embrey (1999)

Infants are particularly susceptible to pharyngitis, gastroenteritis, pneumonia, acute hemorrhagic cystitis and hepatitis associated with adenovirus infection. Ad7 has been shown to cause particularly severe respiratory infections in children. One study documented a 34% mortality ratio from pneumonia and necrotizing bronchiolitis caused by Ad7 (Murtagh et al. 1993). In addition, fulminant hepatitis, caused by adenovirus can be severe in pre-term infants and immunocompromised children (Krilov et al. 1990; Michaels et al. 1992; Munoz et al. 1998; Wang et al. 2003). There have also been reports of adenovirus infection in neonates resulting in serious or fatal pneumonia and disseminated disease (Abzug and Levin 1991; Foy 1997). Evidence from case studies suggests that children may acquire infection from their mothers during birth. Fatal disseminated neonatal infections with Ad3, Ad7, Ad21 and Ad30 have been reported (Abzug and Levin 1991).

It has been speculated that the role of adenoviruses in childhood infections, in developing countries (and also perhaps developed countries), has been underestimated because of the number of asymptomatic infections (Butler et al. 1992). Immunocompromised children are more likely to show clinical symptoms, and malnourished children in developing countries may be predisposed to adenovirus infections. One study investigated the deaths of three children who died from complications associated with concomitant adenovirus and cytomegalovirus infections (Butler et al. 1992).

3.2 Impact on the Immunocompromised

The severity of disease resulting from an adenovirus infection depends on the host's immune system status. Therefore, adenovirus infections in immunocompromised hosts [including HIV (human immunodeficiency virus)-infected patients and transplant recipients] have been well documented (Akiyama et al. 2001; Ambinder et al. 1986; Baldwin et al. 2000; Blohme et al. 1992; Durepaire et al. 1995; Hierholzer 1992; Kampmann et al. 2005; Krilov 2005; Pham et al. 2003; Shields et al. 1985; Webb et al. 1987; Zahradnik et al. 1980). Although non-enteric adenovirus infections usually produce moderate disease in people with normal immune systems, the immunocompromised are at higher risk for serious and possibly fatal disseminated disease.

Though adenovirus infection may result in mild or asymptomatic infections in the immunocompromised (Cox et al. 1994; Khoo et al. 1995), the virus can disseminate into any body system and cause pneumonitis, meningoencephalitis, hepatitis (especially in liver and bone marrow transplant patients) (Bertheau et al. 1996; Saad et al. 1997), and hemorrhagic cystitis (especially in kidney transplant patients) (Foy 1997). The disease may progress to death (Hierholzer 1992) (Table 3). The enteric adenoviruses are rarely isolated from immunocompromised patients with gastroenteritis or diarrhea (Durepaire et al.1995; Khoo et al. 1995), and are generally not associated with serious illness in the immunocompromised; thus, there are no reports of EAds causing chronic sequelae in these individuals.

Hierholzer (1992) found that the different immunocompromised groups tend to be stricken with certain species of the virus. Child SCIDS (severe combined immunodeficiency syndrome) patients were susceptible to serotypes Ad1, Ad7 and Ad31; children with bone marrow transplants were susceptible to the serotypes in subgenera A, B, C and E; adults with kidney transplants were infected with mostly species from subgenus B, especially Ad11, Ad34 and Ad35; and, AIDS patients were documented with infections by all species from all subgenera. The most recently identified species from subgenus D (Ad42 through Ad49 and Ad51), as well as serotype Ad50 (subspecies B1), have only been found in AIDS patients (de Jong et al. 1999; Foy 1997).

3.2.1 Bone Marrow Transplant Patients

Bone marrow transplant patients are at high risk for adenovirus infections. After transplantation, adenovirus isolation rates have been reported as high as 21% in patients (Baldwin et al. 2000). Of those contracting adenovirus, a case-fatality rate of 60% has been reported (Hierholzer 1992). Common syndromes include late-onset hemorrhagic cystitis, diarrhea, pneumonitis, and liver failure (Akiyama et al. 2001;

Hale et al. 1999; Kang et al. 2002). Adenovirus serotypes isolated in bone marrow transplant patients with various syndromes include Ad1, Ad2, Ad5, Ad7, Ad11, Ad34 and Ad35 (Baldwin et al. 2000; Shields et al. 1985; Webb et al. 1987). There is therefore no specific serotype associated with these patients.

3.2.2 AIDS Patients

Adenovirus serotypes commonly isolated from AIDS patients are Ad11, Ad16, Ad21, Ad34 and Ad35 (Hierholzer et al. 1988). It has been estimated that 12% of AIDS patients become infected with adenoviruses and 45% of these infections terminate in death within 2 mon (Hierholzer 1992). Illnesses associated with adenovirus infections and deaths, in AIDS patients, include hepatitis, gastroenteritis, respiratory disease, and diseases of the central nervous system.

4 Waterborne Disease Outbreaks Associated with Adenoviruses

Adenoviruses have been responsible for numerous outbreaks within facilities for: children (e.g., d-care centers, schools, orphanages and camps) (Chiba et al. 1983; McMinn et al. 1991; Payne et al. 1984; Van et al. 1992), hospitals (Brummitt et al. 1988; Colon 1991; de Silva et al. 1989; Koo et al. 1989; McMinn et al. 1991), health care centers (Krajden et al. 1990), and among military personnel (Colon 1991; Dingle and Langmuir 1968; Meiklejohn 1983). Table 5 shows some of the common sources of adenovirus illness outbreaks. Because all serotypes of adenovirus (besides enteric alone) are excreted in feces, contaminated water could be a source of exposure for any type, either through ingestion, inhalation or by direct contact with the eyes. No outbreaks have been associated with food, whereas two outbreaks have been associated with drinking water (Divizia et al. 2004; Kukkula et al. 1997). No water-related outbreaks of enteric adenovirus have been reported.

Table 5 Adenovirus outbreaks from contact with recreational water

Serotype	Source	Population	Disease	Attack rate (%)	Reference
Ad3	Swimming pool	Swim team (ages 8–10 yr)	PCF	67.0	Foy et al. (1968)
		Swim team (ages 10–18 yr)		65.0	
Ad7	Swimming pool	Family open swim	Conjunctivitis	33.3	Caldwell et al. (1974)
Ad4	Swimming pool	Swim team	PCF	52.5	D'Angelo et al. (1979)
Ad3	Swimming pool	Community (ages 1–47 yr)	Conjunctivitis	32.0	Martone et al. (1980)

PCF pharyngoconjunctival feverw

4.1 Recreational Outbreaks

Contact with recreational water has been associated with numerous adenovirus outbreaks over the yr. Adenoviruses are the most reported cause of swimming pool outbreaks associated with viruses (Gerba and Enriquez 1997). Many outbreaks of PCF from non-enteric adenoviruses have come from people swimming in pools and lakes. Papers published in the 1920s described a syndrome of pharyngitis, conjunctivitis and fever (the hallmarks of PCF) related to swimming (Foy 1997). After virus culture techniques became available in the 1950s, these types of swimming pool outbreaks were traced to adenoviruses. Ad7 and Ad3 (the cause of PCF infections) were cultured from the throats of infected people in pool-related outbreaks coinciding with decreases in water chlorine levels (Martone et al. 1980; Turner et al. 1987). Published accounts have confirmed a link with swimming pool outbreaks and the detection of adenoviruses in pool water. Ad3, Ad4, Ad7 and Ad14 have been associated with outbreaks in swimming pools (Gerba and Enriquez 1997; van Heerden et al. 2005a). Ad4 was detected in the water of a Georgia pool after 72 people became ill (D'Angelo et al. 1979). More recently, Greek researchers used polymerase chain reaction (PCR) to detect adenovirus in pool water after 80 swimmers developed PCF (Papapetropoulou and Vantarakis 1998). It is clear that non- or inadequately-disinfected recreational water is a source of adenovirus infection in swimmers. A routine monitoring of chlorinated swimming pools in South Africa demonstrated the presence of adenovirus by PCR in 26 of 93 (15.4%) samples (van Heerden et al. 2005a). Although the detection method did not assess virus viability, it did demonstrate the widespread occurrence of adenoviruses in swimming pools.

4.2 Drinking Water Outbreaks

There have been three drinking water outbreaks reported in Europe in which EAds may have been a cause of gastroenteritis (Divizia et al. 2004; Kukkula et al. 1997; Villena et al. 2003). Multiple viral agents were involved and the water had not been adequately disinfected.

5 Occurrence and Survival of Adenoviruses in Water

Limited data have been available on the occurrences of adenoviruses in water. Only since the development of molecular methods for the direct detection of adenoviruses in water, with confirmation tests performed in cell culture, have data become available. Adenoviruses have been isolated from wastewater and river water, often more frequently and at higher concentrations than the enteroviruses (Hurst et al. 1988; Irving and Smith 1981; Pina et al. 1998). Adenoviruses have also been detected in oceans, swimming pools, and shellfish. Positive samples are found yr-round.

5.1 Occurrence in Sewage

Adenoviruses are commonly detected in raw and non-disinfected secondary sewage discharges, although little published data are available for the US. Table 6 shows the occurrences of adenoviruses in sewage and surface waters worldwide. In Spain,

Table 6 Occurrence of adenoviruses in sewage and surface waters

Serotype(s)[a]	Water source	Concentration/ Frequency	Location	Reference
Ad_{NS}	Raw	0–6,350 IU/L (96%)	Australia	Irving and Smith (1981)
	Secondary	0–600 IU/L (88%)		
	Secondary chlorinated	0–1,150 IU/L (71%)		
Ad1–3, Ad5, Ad7, Ad15	Wastewater	70–3200 cytopathogenic units/L	Greece	Krikelis et al. (1985)
Ad2, Ad3, Ad5, Ad6	River	0–25 PFU/L	Japan	Tani et al. (1995)
Ad_{NS}	Wastewater		Ohio	Hurst et al. (1988)
Ad40, Ad41	Surface	49–88%	South Africa	Genthe et al. (1995)
Ad_{NS}	Wastewater	100%	Spain	Puig et al. (1994)
Ad_{NS}	River	100%		
Ad_{NS}	River		Spain	Pina et al. (1998)
Ad_{NS}	Wastewater		Spain	Girones et al. (1995)
Ad_{NS}	River			
Ad40	Surface water	0–2.11 MPN/48%	United States (US)	Chapron et al. (2000)
Ad	Surface water	880–7,500 genomes/L	California	Jiang et al. (2001)
Ad	River	50%	California	Jiang and Chu (2004)
Ad	Surface water	66.70%	South Korea	Lee and Jeong (2004)
Ad	Surface water	12.75%	South Africa	van Heerden et al. 2003
Ad2, Ad40, Ad41	Surface water	22.22%	South Africa	van Heerden et al. (2005b)
Ad	Sewage effluent and river receiving discharge	20%	Germany	Pusch et al. (2005)
Ad	Secondary	Not given	Milwaukee	Sedmak et al. (2005)
Ad	Raw sewage	94%	Norway	Myrmel et al. (2006)
	Secondary	96%		

[a]Ad_{NS} adenovirus, exact type not specified; *IU* infectious unit; *PFU* plaque forming unit; *MPN* most probable number

monthly samples of raw sewage, effluent, river water, and seawater were tested, using nested PCR amplification. Adenovirus was detected in 14 of 15 sewage, 2 of 3 effluent, 15 of 23 river water, and 7 of 9 seawater samples.

Samples that were positive for enterovirus or hepatitis A were also positive for adenovirus, but there was no correlation between the fecal coliform level and adenovirus occurrence (Pina et al. 1998). In Greece, 36 samples of effluent were tested over a 15-mon period using cell culture. Adenovirus was detected in all samples, with concentrations ranging from 70–3,200 cytopathic units/L. Ad1, Ad2, Ad3, Ad5, Ad7 and Ad15 were detected (Krikelis et al. 1985). In Australia, raw sewage, primary effluent, and secondary effluent were sampled over a yr using cell culture; 25 of 26 raw sewage, 23 of 26 primary effluent and 23 of 26 secondary effluent samples were positive for adenovirus. The mean concentrations in sewage, primary effluent and secondary effluent were 1,950, 1,350 and 250 infectious units/L, respectively. Enteroviruses were removed by activated sludge to a greater extent than was adenoviruses (Irving and Smith 1981).

5.2 *Occurrence in Surface Waters*

Both respiratory and enteric adenoviruses have been isolated from surface waters worldwide. Nevertheless, survey data are limited in the US. An evaluation of 29 surface water samples, collected as part of the Information Collection Rule, yielded 38% positive for infectious Ad40 and Ad41 (Chapron et al. 2000). The concentration of adenovirus 40/41 ranged from 1.03 to 3.23 per 100 L. In this study, adenoviruses were more common in surface waters than enteroviruses and astroviruses. Similarly, when comparative studies were conducted, adenoviruses usually outnumber enteroviruses in surface waters. In Japan, weekly samples of urban river water were tested for reovirus, enteroviruses and adenoviruses, using cell culture, over a period of 5 yr. Levels of adenovirus were low when compared to other viruses, but these levels were consistently detected over the study period (samples ranged from 0 to 25 plaque forming units/L). Ad2, Ad3, Ad5 and Ad6 were the most prevalent forms found (Tani et al. 1995).

5.3 *Occurrence in Groundwater*

To date, there appear to have been no attempts to determine the occurrences of adenoviruses in groundwater, although other enteric viruses have been detected in several studies in the US (Abbaszadegan et al. 2003). Nonetheless, adenovirus was one of the probable causes of a drinking water outbreak in Finland. This indicated that adenovirus might be present in sewage-contaminated groundwater (Kukkula et al. 1997).

5.4 Occurrence in Drinking Water

Infectious adenoviruses have been detected in conventionally treated and disinfected drinking water, in Africa and Asia, using genome detection with PCR in cell culture (Lee and Jeong 2004; van Heerden et al. 2003) (Table 7). In both of these studies, adenoviruses were commonly detected in raw, untreated surface water. In one study, adenoviruses were found in 4.4% of the finished drinking water samples that met the current acceptable bacteriological standards. In the other study, adenoviruses were detected at concentrations ranging from 0 to 0.9 MPN (most probable number)/100 L. Van Heerden et al. (2003) noted that none of the adenoviruses growing in cell culture produced cytopathic effects (CPE). It was reported that all drinking water samples had fewer than 100 colony-forming units of heterotrophic plate count bacteria and no detectable coliform bacteria/100 mL. No studies on adenoviruses, in finished water supplies in the US, have been reported.

5.5 Survival in the Environment

Limited data suggest that adenoviruses survive longer in water than enteroviruses and hepatitis A virus (Enriquez et al. 1995). Adenoviruses also exhibit greater thermal stability than enteroviruses. This may explain their longer survival in water (Enriquez 1999). They are capable of surviving for mon in water, especially at low temperatures. The double-stranded DNA that comprises the genome of the virus may provide more stability in the environment. In addition, adenoviruses may use host cell repair enzymes to repair damaged DNA. This may also prolong their survival in the environment and enhance their resistance to inactivation by ultraviolet (UV) light (Thurston-Enriquez et al. 2003b).

Table 7 Detection of adenoviruses in drinking water

Location	Type of treatment	Method of detection	Concentration/ Frequency of isolation	Reference
South Africa	Not specified	Gene probe for Ad 40 and 41	58% Ad40, 47% Ad41	Genthe et al. (1995)
Korea	Conventional with disinfection	Cell culture and confirmation by PCR	46% Ad	Lee and Jeong (2004)
South Africa	Conventional with chlorination	Integrated cell culture and PCR	4.41% Ad	van Heerden et al. 2003
South Africa	Conventional with chlorination	PCR	5.32% Ad2, Ad40, Ad41	van Heerden et al. (2005b)

6 Economic Impact of Adenovirus Infections

Because adenoviruses are common causes of respiratory disease and gastroenteritis in children (only second to rotavirus gastroenteritis), the economic costs associated with adenovirus infections are not negligible. Although no data on the costs associated with adenovirus gastroenteritis could be found, in 1993, Smith et al. (1995) estimated the direct medical and indirect costs of non-hospitalized rotavirus gastroenteritis at $419 per case, and hospitalized rotavirus gastroenteritis at $3,940 per case. It was estimated that 12,870 cases of acute respiratory infections, caused by adenoviruses among new military recruits in the US, would require hospitalization each yr at an annual cost of $26.4 million (Howell et al. 1998). Outpatient medical costs were estimated at $51 and in-patient medical costs at $1,612. Total direct and indirect costs were estimated at $2,134 per patient.

Eye infections with adenoviruses are common among all age groups. The medical and indirect costs associated with an outbreak of adenoviral conjunctivitis among staff at a long-term care facility were estimated by Pierdnoir et al. (2002) at $722 per person. Staff absenteeism was the most costly aspect of the infection (58.2% of the total cost).

7 Risk Assessment

7.1 Dose-Response

The only dose-response data that exist for adenoviruses are for the respiratory adenovirus, Ad4 (Couch et al. 1966). Haas et al. (1993) determined that the dose-response data were best described by the exponential model:

$$P_i = 1 - \exp(-rN)$$

where, P_i represents the probability of becoming infected, N represents the number of organisms inhaled or ingested, and r is the constant describing the dose-response. The analysis showed that r = 0.4172.

7.2 Previous Risk Assessments

Three previous studies have been conducted to characterize the risk from adenoviruses in drinking and recreational waters. Crabtree et al. (1997) determined the risk of infection, illness and death from adenoviruses in drinking water, using the dose-response data for inhalation of Ad4 (Couch et al. 1966). For data on occurrence, they used the ranges of enteric viruses reported in the literature. Risks of illness and death were determined by multiplying the P_i (probability of infection) by the morbidity rate for waterborne viruses (0.5; Haas et al. 1993) and the reported mortality rate (0.0001; Bennett et al. 1987), respectively. Yearly risks were determined using the following equation:

$$P_{yr} = 1 - (1 - P_i)^{365}$$

The results of their risk analysis are shown in Table 8. They concluded that the risks would exceed the suggested risk recommendation of 1×10^{-4} per yr for drinking water (Regli et al. 1991), even at an adenovirus concentration of 1 per 1,000 L, if the daily consumption of tap water were 2 L/d.

Van Heerden et al. (2005c), using the same exposure and dose response as Crabtree et al. (1997), performed a risk assessment for their data on the occurrence of adenoviruses in conventionally treated drinking water and recreational waters (lake and river) in South Africa. The mean concentration of adenoviruses in the drinking water of the two plants studied was 1.40 and 2.45 adenoviruses per 10,000 L. In the recreational waters studied, the mean concentrations were 54.6 and 9.97 adenovirus per 10,000 L. To obtain these values, a random distribution of the viruses within and between samples was assumed to be described by a Poisson distribution. The assay method used only detected cell culture-infectious virus. The annual risk of infection for drinking water was calculated as 1.01×10^{-1} and 1.71×10^{-1} for the two drinking water supplies. The daily risk of infection was calculated as 1.71×10^{-4} and 3.12×10^{-5} for recreational water.

The risk of adenovirus infection from chlorinated swimming pools has also been assessed. Ninety-two 1-L samples were collected from three swimming pools and assayed for adenoviruses using PCR. Overall, adenoviruses could be detected in 16% of the samples. The authors made the following assumptions: 50% of the adenoviruses were infectious; the viruses had a random distribution in the water; the efficiency of the recovery method was 40%; the concentration of infectious virus in the water was 0.113–0.236 per L; the risk of illness was 50%; and, the swimmers ingested 30 mL. The results indicated a daily risk of adenovirus infection of 1.93×10^{-3} to 3.69×10^{-3}. The total and fecal coliform numbers were within the expected range for coliforms and fecal coliforms (0/100 mL). In this study, residual free chlorine levels (<0.1 mg/L) were detected in some of the swimming pool waters.

Table 8 Risks associated with adenoviruses in drinking and recreational waters

	Yearly risk for drinking water[a]		Daily risk for swimming[b]	
Risk	(1 IU/1,000 L)	(1 IU/100 L)	(0.118 MPN/100 L)	(12.8 MPN/100 L)
Infection	2.63×10^{-1}	9.52×10^{-1}	1.48×10^{-5}	1.60×10^{-3}
Illness	1.41×10^{-1}	7.81×10^{-1}	7.38×10^{-6}	8.00×10^{-4}
Death	1.52×10^{-5}	1.52×10^{-4}	7.38×10^{-10}	8.00×10^{-8}

IU infectious unit
[a]Assumes the general population consumes 2 L/d
[b]30 mL ingested

7.3 Risks for Drinking Water

Table 9 shows the daily and annual risks of infection associated with exposure to varying levels of adenoviruses per 100 L. The calculated risks assume a 2 L/d exposure and utilize the exponential model with $r = 0.4172$ (Haas et al. 1993).

A guideline of acceptable risk for drinking water of 1×10^{-4} has been suggested (Regli et al. 1991). At every exposure included in Table 9, calculated annual risks exceed this recommendation. If it is assumed that these adenovirus concentrations are detected in source water, the amount of water treatment required to reduce the risk to meet the EPA recommendation, can be determined. Figure 2 shows the relationship of adenovirus concentration and the $\log_{10}$ reduction through treatment needed to meet the 1×10^{-4} annual risk of infection goal.

If current guidelines for adenovirus removal, using conventional disinfection treatments, are capable of removing 4-$\log_{10}$ of adenoviruses, then surface water concentrations should not exceed ~0.5 per 100 L (Fig. 2). Chapron et al. (2000) detected adenoviruses at 4 of 29 surface drinking-water treatment sites that were involved in the Information Collection Rule data collection. The concentration of

Table 9 Daily and annual risks of infection associated with exposure to waterborne adenoviruses

No. of adenoviruses per 100 L	Daily risk of infection	Annual risk of infection
0.01	8.34×10^{-5}	2.99×10^{-2}
0.1	8.34×10^{-4}	2.63×10^{-1}
1	8.31×10^{-3}	9.52×10^{-1}
10	8.01×10^{-2}	9.99×10^{-1}
100	5.65×10^{-1}	9.99×10^{-1}

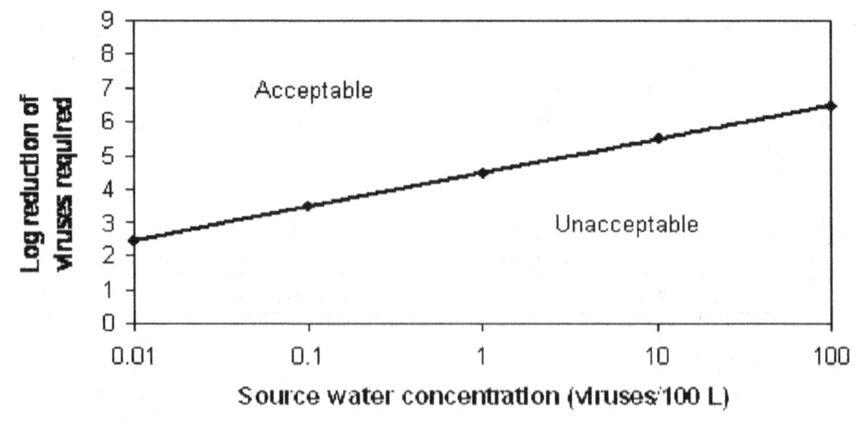

Fig. 2 $\log_{10}$ reduction of adenoviruses in source water required to achieve acceptable finished water (risk of yearly infection of $<10^{-4}$)

adenoviruses ranged from 1.16 to 2.11 MPN/100 L, suggesting that additional treatment may be needed at these sites. The effectiveness of conventional treatment for adenoviruses is not known. Existing data would suggest that chlorine doses, used for the control of other enteric viruses, are more effective against adenovirus, and greater than a 4-$\log_{10}$ removal may occur by conventional treatment followed by chlorination.

No information is available on the occurrence of adenoviruses in groundwater. However, several studies have documented the occurrence of enteroviruses, and other enteric viruses in groundwater used as a source for drinking water in the US (Abbaszadegan et al. 2003; Borchardt et al. 2003, 2004; Fout et al. 2003). Abbaszadegan et al. (2003) detected viruses in the BGM (Buffalo green monkey) cell line by CPE in 22 of 529 samples from utility water supply wells, with an MPN viral concentration between 0.09 and 1.86 MPN/100 L. If adenoviruses occur in similar concentrations, then a 4.0–4.5-$\log_{10}$ removal would be needed to achieve a yearly risk of 1×10^{-4}.

8 Removal of Adenoviruses by Water Treatment

No published studies on the removal of adenoviruses by conventional treatment, or other physical-chemical treatment processes of drinking water are available. Although adenoviruses have been isolated from conventionally treated drinking water in developing countries, no such studies have been performed in the US.

9 Disinfection

Only limited studies have been performed on the sensitivity of adenoviruses to the disinfectants commonly used in water treatment (Thurston-Enriquez et al. 2003a, b). These studies suggest that adenoviruses are of equal or greater sensitivity to oxidizing disinfectants; however, because of its doubled-stranded DNA genome, they appear to be the most UV light - resistant waterborne pathogen known. Table 10 lists the Ct (concentration **x** time) values for various disinfectants used in the treatment of water.

Among the non-enteric adenoviruses, sensitivities to chlorine appear to be similar. The EAd40 appears to be very sensitive. Gerba et al. (2002) estimated the Ct for a 99.99% reduction at a temperature of 2 C to be 2.4 for Ad3, 6.25 for Ad7 and 6.75 for Ad12 (from the work of Liu et al. 1971). Thurston-Enriquez et al. (2003a) found that the Ct for a 99% inactivation of Ad40 was less than that of poliovirus and the other adenoviruses, in buffered demand-free water at 5 C at pH 6.0–8.0. The Ct for a 2-log inactivation (99%) was 13.7 times greater in groundwater supplied as drinking water. Additional work conducted as part of the study confirmed that adenoviruses are more resistant to chlorine in groundwater and surface waters

Table 10 Ct (concentration × time) values for various disinfectants

Serotype	Disinfectant	Water type	Temperature (C)	pH	Estimated Ct_{99} (mg/min)	Reference
Ad3	Chlorine	PEW	2	7.8	2.40	Liu et al. (1971)
Ad7A	Chlorine	PEW	2	7.8	6.25	Liu et al. (1971)
Ad11	Chlorine	PEW	2	7.8	6.75	Liu et al. (1971)
Ad5	Chlorine	Tap	25	7.0	6.00	Abad et al. (1994)
Ad40	Chlorine	CDF	5	7.0	0.03	Thurston-Enriquez et al. (2003a)
Ad40	Chlorine	CDF	5	8.0	0.11	Thurston-Enriquez et al. (2003a)
Ad40	Chlorine dioxide	CDF	5	7.0	0.28	Thurston-Enriquez et al. (2005a)
Ad40	Ozone	ODF	5–7	7.0	0.01	Thurston-Enriquez et al. (2005b)

PEW potomac estuarine water, *CDF* chlorine demand free, *ODF* ozone demand free

receiving conventional treatment; however, the Ct time is still within values recommended for the treatment of surface waters.

Only the inactivation of Ad40 by ozone has been reported to date (Thurston-Enriquez et al. 2005b). The Ct for a 99% inactivation of Ad40 by ozone is less than 0.01 (pH 7.0, 5 C; Thurston-Enriquez et al. 2005b). Thus, adenoviruses are more sensitive to ozone than are the enteroviruses (Gerba et al. 2003). Ad40 also appears to be very sensitive to inactivation by chlorine dioxide, with a Ct of 0.28 for a 99% reduction at pH 7.0 and 5 C (Thurston-Enriquez et al. 2005a).

Respiratory and enteric adenoviruses are considerably more resistant to UV light radiation than are other waterborne enteric pathogens (Gerba et al. 2002, 2003; Roessler and Severin 1996; Yates et al. 2006). They are also more resistant to UV light disinfection than waterborne viruses with single- and double-stranded RNA genomes (Meng and Gerba 1996). The adenovirus genome is comprised of double-stranded DNA. This allows the virus to use host cell repair enzymes to repair damage in the DNA caused by UV light (Day et al. 1975).

Thurston-Enriquez et al. (2003b) reported a value of 23 and 87 mJ/cm^2 for a 1- and 3-log inactivation of the MS-2 virus, respectively. MS-2 was reduced by 3.1-log_{10} at a dose of 90 mJ/cm^2, as interpolated from the linear regression line. This is greater than the inactivation rate of adenoviruses at the same dose (Table 11). Adenoviruses have a double-stranded DNA genome, but only one strand of the nucleic acid may be damaged during UV light disinfection. The undamaged strand may then serve as a template for host cell repair enzymes (Day et al. 1975; Day 1993). The presence of host cell repair enzymes enable DNA viruses to repair the damage caused by UV light.

In one study of UV light-exposed adenoviruses, two different continuous cell lines – PLC/PRF/5 and HeLa cells – were used to determine if they would provide different results. An analysis of variance (ANOVA) test of the results indicated that the differences between log_{10} (N/N_0) of experiments assayed on PLC and HeLa

Table 11 Predicted dose requirements (mJ/cm^2) for $\log_{10}$ inactivation using ultraviolet light

Virus	90%	99%	99.90%	99.99%
Ad6	38.5	76.9	115.4	153.8
Ad1	34.5	68.9	103.4	137.9
MS-2	28.9	57.9	86.9	115.9
Ad2[a]	40.0	78.0	119.0	160.0
Polio-1[a]	8.0	15.5	23.0	31.0
Ad40[b]	54.3	109.0	167.0	226.0

[a]Gerba et al. (2002)
[b]Thurston-Enriquez et al. (2003b)

cells were not statistically significant. Multiple freeze-thaw cycles of virus stocks have been speculated to damage the viral capsid, making them more susceptible to UV radiation (Gerba et al. 2002). There was not a significant difference between the inactivation rate of Ad1 and Ad6 after one freeze-thaw; nevertheless, the slopes of the linear regression lines of Ad6 after one freeze-thaw and four freeze-thaws showed a significant difference. This indicates that Ad6 was more susceptible to UV light after four freeze-thaws at a dose of 120 mJ/cm^2. There was no statistically significant difference in the $\log_{10}$ reduction of Ad5 and Ad6, stored at two dissimilar temperatures, for up to 4 week before UV light exposure. The exception was Ad1 that had a higher inactivation rate after 4 week of storage at 25 C ($p = 0.02$).

Gerba et al. (2002) found that the doses required to achieve 90–99.99% inactivation of Ad2 were very similar to the ones required for Ad6, and slightly higher than the doses required for Ad1 (Table 11). Thurston-Enriquez et al. (2003b) found that enteric Ad40 was significantly more resistant to UV light than the respiratory adenoviruses. Meng and Gerba (1996) reported 30 and 124 mJ/cm^2 for a 1- and 4-$\log_{10}$ inactivation of Ad40 (frozen and thawed five times prior to exposure). This is very similar to the reductions in Ad6 after four freeze-thaw cycles (29 and 117.6 mJ/cm^2). Also, the reported 3.3-$\log_{10}$ reduction of Ad40 at 90 mJ/cm^2 was almost identical to Ad6 (Table 11).

Yates et al. (2006) recently reviewed the state of knowledge on adenovirus inactivation by UV light, and concluded that all existing data suggest that delivering the dose in a UV reactor (while allowing for uncertainties in reactor validation testing) may require UV doses of up to 200–300 mJ/cm^2 for a 4-$\log_{10}$ virus inactivation for a given UV reactor.

10 Data Gaps

Although adenoviruses have long been recognized as a cause of waterborne illnesses among bathers, they have received less study than other waterborne viruses, with regard to their fate and transport in the environment. Although we know that they occur in greater numbers, in sewage and sewage-polluted surface waters, than do other enteric viruses, we know little concerning the effectiveness of conventional sewage and drinking water treatment for their removal. The data available are usually limited to one or a few serotypes.

Quantitative data on adenovirus occurrence in surface and groundwater would aid in determining the amount of treatment needed to meet treatment goals. Data on conventional water treatment would allow better assessment of the amount of disinfectant needed. Information on adenovirus transport through soil would also assist in understanding the potential adenovirus has for contaminating of groundwater.

Adenoviruses have been shown in laboratory cell cultures to use host repair enzymes to repair UV light damage; however, there is insufficient information to clearly determine the relevance of such repair for human consumption of water that contains UV light-irradiated adenoviruses (Yates et al. 2006). Animal or other studies could be conducted to define the significance of this phenomenon and its importance to UV light disinfection.

Current dose-response data are only available for the inhalation route. It is possible that the dose-response for enteric adenoviruses by ingestion is different. Studies in animals could provide better data on dose-response for adenoviruses via the oral route. Dose-response studies in immunocompromised animals could also aid in determining if the dose-response is different from responses in healthy individuals.

11 Summary

Adenoviruses are associated with numerous disease outbreaks, particularly those involving d-cares, schools, children's camps, hospitals and other health care centers, and military settings. In addition, adenoviruses have been responsible for many recreational water outbreaks, including a greater number of swimming pool outbreaks than any other waterborne virus (Gerba and Enriquez 1997). Two drinking water outbreaks have been documented for adenovirus (Divizia et al. 2004; Kukkula et al. 1997) but none for food. Of the 51 known adenovirus serotypes, one third are associated with human disease, while other infections are asymptomatic. Human diseases associated with adenovirus infections include gastroenteritis, respiratory infections, eye infections, acute hemorrhagic cystitis, and meningoencephalitis (Table 2). Children and the immunocompromised are more severely impacted by adenovirus infections. Subsequently, adenovirus is included in the EPA's Drinking Water Contaminant Candidate List (CCL), which is a list of unregulated contaminants found in public water systems that may pose a risk to public health (National Research Council 1999).

Adenoviruses have been detected in various waters worldwide including wastewater, river water, oceans, and swimming pools (Hurst et al. 1988; Irving and Smith 1981; Pina et al. 1998). Adenoviruses typically outnumber the enteroviruses, when both are detected in surface waters. Chapron et al. (2000) found that 38% of 29 surface water samples were positive for infectious Ad40 and Ad41. Data are lacking, regarding the occurrence of adenovirus in water in the US, particularly for groundwater and drinking water. Studies have shown, however, that adenoviruses survive longer in water than enteroviruses and hepatitis A virus (Enriquez et al. 1995), which may be due to their double-stranded DNA.

Risk assessments have been conducted on waterborne adenovirus (Crabtree et al. 1997; van Heerden et al. 2005c). Using dose-response data for inhalation from Couch et al. (1966), human health risks of infection, illness and death have been determined for various adenovirus exposures. Crabtree et al. (1997) conclude that, even at an adenovirus concentration of 1 per 1,000 L of drinking water, annual risks of infection exceed the suggested risk recommendation of 1×10^{-4} per yr (Regli et al. 1991) (Table 8). Using the same exposure and dose-response assumptions, van Heerden et al. (2005c) determined annual risks of infection to be $1–1.7 \times 10^{-1}$ for two drinking water samples from South Africa containing 1.40 and 2.45 adenoviruses per 10,000 L, respectively. This present study estimated annual risks of infection associated with varying levels of adenoviruses per 100 L (Table 9). By assuming a 2 L/d exposure and utilizing the exponential model at $r = 0.4172$ (Haas et al. 1993), yearly risks exceed the risk recommendation of 1×10^{-4} at every exposure level.

There are limited data regarding the removal of adenoviruses by conventional water treatment or other physical-chemical treatment processes, but studies do suggest that adenoviruses are of equal or greater sensitivity to oxidizing disinfectants, when compared to other waterborne viruses (the most resistant to ultraviolet light). Data suggest that the chlorine doses applied to control other waterborne viruses are more effective against adenovirus, resulting in a greater than 4-$\log_{10}$ removal of adenoviruses by conventional treatment and chlorination. If treatment can achieve a 4-$\log_{10}$ removal of adenoviruses, then, based on the risk levels presented in Table 9, surface water concentrations should not exceed 0.5 adenoviruses per 100 L (Fig. 2). More data are needed regarding the occurrence of adenovirus in groundwater and drinking water, the effectiveness of water treatment against adenovirus, and the human-virus dose-response relationship to fully understand the role of adenovirus as a waterborne public health threat.

References

Abad FX, Pinto RM, Diez JM, Bosch A (1994) Disinfection of human enteric viruses in water by copper and silver in combination with low levels of chlorine. Appl Environ Microbiol 60:2377–2383.

Abbaszadegan M, Lechevallier M, Gerba C (2003) Occurrence of viruses in US groundwaters. J Am Water Works Assoc 95:107–120.

Abzug MJ, Levin MJ (1991) Neonatal adenovirus infection: four patients and review of the literature. Pediatrics 87:890–896.

Adrian T, Wig R, Richter J (1987) Gastroenteritis in infants associated with a genome type of adenovirus 31 and with a combined rotavirus and adenovirus 31 infection. Eur J Pediatr 146:38–40.

Akiyama H, Kurosu T, Sakashita C, Inoue T, Mori S, Ohashi K, Tanikawa S, Sakamaki H, Onozawa Y, Chen Q, Zheng H, Kitamura T (2001) Adenovirus is a key pathogen in hemorrhagic cystitis associated with bone marrow transplantation. Clin Inf Dis 32:1325–1330.

Allard A, Girones R, Juto P, Wadell G (1991) Polymerase chain reaction for detection of adenoviruses in stool samples. J Clin Microbiol 29:2683.

Ambinder RF, Burns W, Forman M, Charache P, Arthur R, Beschorner W, Santos G, Saral R (1986) Hemorrhagic cystitis associated with adenovirus infection in bone marrow transplantation. Arch Intern Med 146:1400–1401.

Atkinson RL, Dhurandhar NV, Allison DB, Bowen RL, Israel BA, Albu JB, Augustus AS (2005) Human adenovirus-36 is associated with increased body weight and paradoxical reduction of serum lipids. Int J Obes 29:281–286.

Baldwin A, Kingman H, Darville M, Foot AB, Grier D, Cornish JM, Goulden N, Oakhill A, Pamphilon DH, Steward CG, Marks DI (2000) Outcome and clinical course of 100 patients with adenovirus infection following bone marrow transplantation. Bone Marrow Transplant 26:1333–1338.

Bates PR, Bailey AS, Wood DJ, Morris DJ, Couriel JM (1993) Comparative epidemiology of rotavirus, subgenus F (types 40 and 41) adenovirus and astrovirus gastroenteritis in children. J Med Virol 39:224–228.

Baum SG (2000) Adenovirus. In: Mandell L, Bennett JE, Dolin R (eds.) Principles and Practice of Infectious Diseases. Churchill Livingstone, Philadelphia, PA, pp. 1624–1630.

Bennett JV, Homberg SD, Rogers MF, Solomon SL (1987) Infectious and parasitic diseases. Am J Prevent Med 55:102–114.

Bertheau P, Parquet N, Ferchal F, Gluckman E, Brocheriou C (1996) Fulminant adenovirus hepatitis after allogeneic bone marrow transplantation. Bone Marrow Transplant 17:295–298.

Blohme I, Nyberg G, Jeansson S, Svalander C (1992) Adenovirus infection in a renal transplant patient. Transplant Proc 24:295.

Borchardt MA, Bertz PD, Spencer SK, Battigelli DA (2003) Incidence of enteric viruses in groundwater from household wells in Wisconsin. Appl Environ Microbiol 69:1172–1180.

Borchardt MA, Haas NL, Hunt RJ (2004) Vulnerability of drinking-water wells in LaCrosse, Wisconsin, to enteric-virus contamination from surface water contributions. Appl Environ Microbiol 70:5937–5946.

Brandt CD, Kim HW, Vargosko AJ, Jeffries BC, Arrobio JO, Rindge B, Parrott RH, Chanock RM (1969) Infections in 18,000 infants and children in a controlled study of respiratory tract disease. I. Adenovirus pathogenicity in relation to serologic type and illness syndrome. Am J Epidemiol 90:484–500.

Brown M (1990) Laboratory identification of adenoviruses associated with gastroenteritis in Canada from 1983 to 1986. J Clin Microbiol 28:1525–1529.

Brummitt CF, Cherrington JM, Katzenstein DA, Juni BA, van Drunen N, Edelman C, Rhame FS, Jordan MC (1988) Nosocomial adenovirus infections: molecular epidemiology of an outbreak due to adenovirus 3a. J Infect Dis 158:423–432.

Bryden AS, Curry A, Cotterill H, Chesworth C, Sharp I, Wood SR (1997) Adenovirus-associated gastro-enteritis in the north-west of England: 1991–1994. Brit J Biomed Sci 54:273–277.

Buescher EL (1967) Respiratory disease and the adenoviruses. Med Clin North Am 51:769–779.

Butler T, Dunn D, Colmer J (1992) Concomitant intestinal adenovirus infection and pulmonary cytomegalovirus infection in children causing fatal enteritis and pneumonia. Trans R Soc Trop Med Hyg 86:298–300.

Caldwell GG, Lindsey NJ, Wulff H, Donnelly DD, Bohl FN (1974) Epidemic adenovirus type 7 acute conjunctivitis in swimmers. Am J Epidemiol 99:230–234.

CDC (Centers for Disease Control and Prevention) (1998) Civilian outbreak of adenovirus acute respiratory disease-South Dakota, 1997. MMWR 47:567–570.

CDC (Centers for Disease Control and Prevention) (2001) Two fatal cases of adenovirus-related illness in previously healthy young adults-Illinois, 2000. MMWR 50:553–555.

Chapron CD, Ballester NA, Fontaine JH, Frades CN, Margolin AB (2000) Detection of astroviruses, enteroviruses, and adenovirus types 40 and 41 in surface water collected and evaluated by the information collection rule and a integrated cell culture-nested PCR procedure. Appl Environ Microbiol 66:2520–2525.

Chiba S, Nakata S, Nakamura I, Taniguchi K, Urasawa S, Fujinaga K, Nakao T (1983) Outbreak of infantile gastroenteritis due to type 40 adenovirus. Lancet 2:954–957.

Colon LE (1991) Keratoconjunctivitis due to adenovirus type 8: report on a large outbreak. Ann Ophthalmol 23:63–65.

Cooney MK, Hall CE, Fox JP (1972) The Seattle virus watch. 3. Evaluation of isolation methods and summary of infections detected by virus isolations. Am J Epidemiol 96:286–305.

Couch RB, Cate TR, Douglas RG, Jr, Gerone PJ, Knight V (1966) Effect of route of inoculation on experimental respiratory viral disease in volunteers and evidence for airborne transmission. Bacteriol Rev 30:517–529.

Cox GJ, Matsui SM, Lo RS, Hinds M, Bowden RA, Hackman RC, Meyer WG, Mori M, Tarr PI, Oshiro LS, Ludert JE, Meyers JD, McDonald GB (1994) Etiology and outcome of diarrhea after marrow transplantation: a prospective study. Gastroenterology 107:1398–1407.

Crabtree KD, Gerba CP, Rose JB, Haas CN (1997) Waterborne adenovirus: a risk assessment. Water Sci Technol 35:1–6.

Cruz JR, Caceres P, Cano F, Flores J, Bartlett A, Torun B (1990) Adenovirus types 40 and 41 and rotaviruses associated with diarrhea in children from Guatemala. J Clin Microbiol 28:1780–1784.

D'AngeloLJ, HierholzerJC, KeenlysideRA, AndersonLJ, MartoneWJ (1979) Pharyngoconjunctival fever caused by adenovirus type 4: report of a swimming pool-related outbreak with recovery of virus from pool water. J Infect Dis 140:42–47.

D'Angelo LJ, Hierholzer JC, Holman RC, Smith JD (1981) Epidemic keratoconjunctivitis caused by adenovirus type 8: epidemiologic and laboratory aspects of a large outbreak. Am J Epidemiol 113:44–49.

Day RS (1993) Deoxyguanosine reverse inhibition by hydroxyurea of repair of UV-irradiated adeovirus 5. Mutat Res 293:215–223.

Day RS, Giuffrida AS, Dingman CW (1975) Repair by human cell of aenovirus 2 damaged by psoralen plus near ultraviolet light treatment. Mutat Res 33:311–320.

de Jong JC, Wigand R, Kidd AH, Wadell G, Kapsenberg JG, Muzerie CJ, Wermenbol AG, Firtzlaff RG (1993) Candidate adenoviruses 40 and 41: fastidious adenoviruses from human infant stool. J Med Virol 11:215–231.

de Jong JC, Wermenbol AG, Verweij-Uijterwaal MW, Staterus KW, Wertheim-van Dillen P, van Doornum GJ, Khoo SH, Heirhoizer JC (1999) Adenoviruses from human immunodeficiency virus-infected individuals, including two strains that represent new candidate serotypes Ad50 and Ad51 of species B1 and D, respectively. J Clin Microbiol 37:3940–3945.

de Silva LM, Colditz P, Wadell G (1989) Adenovirus type 7 infections in children in New South Wales, Australia. J Med Virol 29:28–32.

Dingle JH, Langmuir AD (1968) Epidemiology of acute, respiratory disease in military recruits. Am Rev Respir Dis 97:1–65.

Divizia M, Gabrieli D, Donia A, Macaluso A, Bosch A, Guiz S, Sanchez G, Villena C, Pinto RM, Palombi L, Buonuomo E, Cenko F, Leno L, Bebeci D, Bino S (2004) Waterborne gastroenteritis outbreak in Albania. Water Sci Technol 50:57–61.

Donelli G, Superti F, Tinari A, Marziano ML, Caione D, Concato C, Menichella D (1993) Viral childhood diarrhoea in Rome: a diagnostic and epidemiological study. New Microbiol 16:215–225.

Dudding BA, Wagner SC, Zeller JA (1972) Fatal pneumonia associated with adenovirus type 7 in three military trainees. N Engl J Med 286:1289–1292.

Durepaire N, Ranger-Rogez S, Gandji JA, Weinbreck P, Rogez JP, Denis F (1995) Enteric prevalence of adenovirus in human immunodeficiency virus seropositive patients. J Med Virol 45:56–60.

Embrey M (1999) Adenovirus in Drinking Water: Literature Summary. Prepared for the Office of Water under Cooperative Agreement #CX8236396-01-0, April 1999..

Enders JF, Bell JA, Dingle JH, Francis T, Hilleman MR, Huebner RJ, Payne AM (1956) Adenoviruses: group name proposed for new respiratory-tract viruses. Science 124:119–120.

Enriquez CE (1999) Adenoviruses. In: American Water Works Association (ed.) Waterborne pathogens. AWWA manual M48. American Water Works Association, Denver, pp. 223–226.

Enriquez CE (2002) Adenoviruses. In: Bitton G (ed.) Encyclopedia of environmental microbiology, Vol. 1. John Wiley & Sons, Inc., New York, NY, pp. 92–100.

Enriquez CE, Hurst CJ, Gerba CP (1995) Survival of the enteric andenoviruses 40 and 41 in tap, sea, and waste water. Wat Res 29:2548–2553.

Evans TN, Brown GC (1963) Congenital anamolies in virus infection. Am J Obstet Gyneco 87:749–761.

Farng NT, Wu KG, Lee YS, Lin YH, Hwang BT (2002) Comparison of clinical characteristics of adenovirus and non-adenovirus pneumonia in children. J Microbiol Immunol Infect 35:37–41.

Fenger TW (1991) Replication of DNA viruses. In: Belshe RB (ed.) Textbook of human virology, 2nd Ed. Mosby Year Book, St Louis, MO, pp. 24–38.

Fout GS, Martinson BC, Moyer MW, Dahling DR (2003) A multiplex reverse transcription-PCR method for detection of human enteric viruses in groundwater. Appl Environ Microbiol 69:3158–3164.

Fox JP, Brandt CD, Wasserman FE, Hall CE, Spigland I, Kogon A, Elveback LR (1969) The virus watch program: a continuing surveillance of viral infections in metropolitan New York families. VI. Observations of adenovirus infections: virus excretion patterns, antibody response, efficiency of surveillance, patterns of infections and relation to illness. Am J Epidemiol 89:25–50.

Fox JP, Hall CE, Cooney MK (1977) The Seattle virus watch VII. Observations of adenovirus infections. Am J Epidemiol 105:362–386.

Foy HM (1997) Adenoviruses. In: Evans AS, Kaslow RA (eds.) Viral infections of humans: epidemiology and control, 4th Ed. Plenum Publishing Corporation, New York, NY, pp. 119–138.

Foy HM, Cooney MK, Halten J (1968) Adenovirus type 3 epidemic associated with intermittent chlorination of a swimming pool. Arch Environ Health 17:795–802.

Freymuth F, Vabret A, Gouarin S, Petitjean J, Charbonneau P, Lehoux P, Galateau-Salle F, Tremolieres F, Carette MF, Mayaud C, Mosnier A, Burnouf L (2004) Epidemiology and diagnosis of respiratory syncitial virus in adults. Rev Mal Respir 21:35–42.

Genthe B, Gericke M, Bateman B, Mjoli N, Kfir R (1995) Detection of enteric adenoviruses in south African water using gene probes. Wat SciTech 31:345–350.

Gerba CP, Enriquez CE (1997) Virus-assoicated outbreaks in swimming pools. In: Denkewicz R, Gerba CP, Hales Q (eds.) Water chemistry and disinfection: swimming pools and spas. National Spa and Pool Institute, Alexandria, pp. 31–45.

Gerba CP, Gramos DM, Nwachuku N (2002) Comparative inactivation of enteroviruses and adenovirus 2 by UV light. Appl Environ Microbiol 68:5167–5169.

Gerba CP, Nwachuku N, Riley KR (2003) Disinfection resistance of waterborne pathogens on the United States Environmental Protection Agency's Contaminant Candidate List (CCL). J Water Supply Res Technol 52:81–94.

Girones R, Puig M, Allard A, Lucena F, Wadell G, Jofre J (1995) Detection of adenovirus and enterovirus by PCR amplification in polluted waters. Water Sci Tech 31:351–357.

Greenway F (2006) Virus-induced obesity. Am J Physiol Regul Integr Comp Physiol 290:R188–R189.

Grimwood K, Carzino R, Barnes GL, Bishop RF (1995) Patients with enteric adenovirus gastroenteritis admitted to an Australian pediatric teaching hospital from 1981 to 1992. J Clin Microbiol 33:131–136.

Gutekunst RR, Heggie AD (1961) Viremia and viruria in adenovirus infections. Detection in patients with rubella or rubelliform illness. N Engl J Med 264:374–378.

Haas CN, Rose JB, Gerba CP, Regli S (1993) Risk assessment of virus in drinking water. Risk Anal 13:545–552.

Hale GA, Heslop HE, Krance RA, Brenner MA, Jayawardene D, Srivastava DK, Patrick CC (1999) Adenovirus infection after pediatric bone marrow transplantation. Bone Marrow Transplant 23:277–282.

Harley D, Harrower B, Lyon M, Dick A (2001) A primary school outbreak of pharyngo-conjunctival fever caused by adenovirus type 3. Commun Dis Intell 25:9–12.

Harsi CM, Rolim DP, Gomes SA, Gilio AE, Stewien KE, Baldacci ER, Candeias JA (1995) Adenovirus genome types isolated from stools of children with gastroenteritis in Sao Paulo, Brazil. J Med Virol 45:127–134.

Hierholzer JC (1992) Adenoviruses in the immunocompromised host. Clin Microbiol Rev 5:262–274.

Hierholzer JC, Wigand R, Anderson LJ, Adrian T, Gold JW (1988) Adenoviruses from patients with AIDS: a plethora of serotypes and a description of five new serotypes of subgenus D (types 43–47). J Infect Dis 158:804–813.

Hilleman MR, Werner JH (1954) Recovery of new agents from patients with acute respiratory illness. Proc Soc Exp Biol Med 85:183–188.

Horwitz MS (2001) Adenoviruses. In: Knipe DM, Howley PM, Griffin DE, Lamb RA, Martin MA, Roizman B, Straus SE (eds.) Fields Virology, 4th Ed. Lippincott Williams & Wilkins, Philadelphia, PA, pp. 2301–2326.

Howell RM, Major R, Nang N, Gaydos CA, Gaydos JC (1998) Prevention of adenovirus acute respiratory disease in army recruits: cost-effectiveness of a military vaccination policy. Am wJ Prev Med 14:168–175.

Hurst CJ, McClellan KA, Benton WH (1988) Comparison of cytopathogenicity, immunofluorescence and *in situ* DNA hybridization as methods for the detection of adenoviruses. Wat Res 22:1547–1552.

Irving LG, Smith FA (1981) One-year survey of enteroviruses, adenoviruses, and reoviruses isolated from effluent at an activated-sludge purification plant. Appl Environ Microbiol 41:51–59.

Jaworowska A, Barylak G (2006) Obesity development associated with viral infections. Postepy Hig Med Dosw 60:227–236.

Jiang S, Chu W (2004) PCR detetion of pathogenic viruses in southern California urban rivers. J Appl Microbiol 97:17–28.

Jiang S, Noble R, Chu W (2001) Human adenoviruses and coliphages in urban runoff-impacted coastal waters of Southern California. Appl Environ Microbiol 67:179–184.

Kampmann B, Cubitt D, Walls T, Naik P, Depala M, Samarasinghe S, Robson D, Hassan A, Rao K, Gaspar H, Davies G, Jones A, Cale C, Gilmour K, Real M, Foo M, Bennett-Rees N, Hewitt A, Amrolia P, Veys P (2005) Improved outcome for children with disseminated adenoviral infection following allogeneic stem cell transplantation. Brit J Haematol 130:595–603.

Kang G, Srivastava A, Pulimood AB, Dennison D, Chandy M (2002) Etiology of diarrhea in patients undergoing allogeneic bone marrow transplantation in South India. Transplantation 73:1247–1251.

Kawai T, Fujiwara T, Aoyama Y, Aizawa Y, Yamada Y (1976) Diffuse interstitial fibrosing pneumonitis and adenovirus infection. Chest 69:692–694.

Keenlyside RA, Hierholzer JC, D'Angelo LJ (1983) Keratoconjunctivitis associated with adenovirus type 37: an extended outbreak in an ophthalmologist's office. J Infect Dis 147:191–198.

Kemp MC, Hierholzer JC, Cabradilla CP, Obavijesti JF (1983) The changing etiology of epidemic keratoconjunctivitis: antigenic and restriction enzyme analysis of adenovirus 19 and 37 isolated over a 10 year period. J Infect Dis 148:29–33.

Khoo SH, Bailey AS, de Jong JC, Mandal BK (1995) Adenovirus infections in human immunodeficiency virus-positive patients: clinical features and molecular epidemiology. J Infect Dis 172:629–637.

Kim K, Yang J, Joo S, Cho Y, Glass RI, Cho YJ (1990) Importance of rotavirus and adenovirus types 40 and 41 in acute gastroenteritis in Korean children. J Clin Microbiol 28:2279–2284.

Koo D, Bouvier B, Wesley M, Courtright P, Reingold A (1989) Epidemic keratoconjunctivitis in a university medical center ophthalmology clinic; need for re-evaluation of the design and disinfection of instruments. Infect Control Hosp Epidemiol 10:547–552.

Kotloff KL, Wasserman SS, Steciak JY, Tall BD, Losonsky GA, Nair P, Morris JG, Jr, Levine MM (1988) Acute diarrhea in Baltimore children attending an outpatient clinic. Pediatr Infect Dis J 7:753–759.

Kowalski RP, Sundar-Raj CV, Romanowski EG, Gordon YJ (2001) The disinfection of contact lenses contaminated with adenovirus. Am J Ophthalmol 132:777–779.

Krajden M, Brown M, Petrasek A, Middleton PJ (1990) Clinical features of adenovirus enteritis: a review of 127 cases. Pediatr Infect Dis J 9:636–641.

Krikelis V, Spyrou N, Markoulatos P, Serie C (1985) Seasonal distribution of enteroviruses and adenoviruses in domestic sewage. Can J Microbiol 31:24–25.

Krilov LR (2005) Adenovirus infections in the immunocompromised host. Pediatr Infect Dis J 24:555–556.

Krilov LR, Rubin LG, Frogel M, Gloster E, Ni K, Kaplan M, Lipson SM (1990) Disseminated adenovirus infection with hepatic necrosis in patients with human immunodeficiency virus infection and other immunodeficiency states. Rev Infect Dis 12:303–307.

Kukkula M, Arstila P, Klossner ML, Maunula L, Bonsdorff CH, Jaatinen P (1997) Waterborne outbreak of viral gastroenteritis. Scand J Infect Dis 29:415–418.

LeBaron CW, Furutan NP, Lew JF, Allen JR, Gouvea V, Moe C, Monroe SS (1990) Viral agents of gastroenteritis. Public health importance and outbreak management. MMWR 39:1–24.

Lee HK, Jeong YS (2004) Comparison of total culturable virus assay and multiplex integrated cell culture-PCR for reliability of waterborne virus detection. Appl Environ Microbiol 70:3632–3636.

Lew JF, Moe CL, Monroe SS, Allen JR, Harrison BM, Forrester BD, Stine SE, Woods PA, Hierholzer JC, Herrmann JE (1991) Astrovirus and adenovirus associated with diarrhea in children in day care settings. J Infect Dis 164:673–678.

Liu C (1991) Adenoviruses. In: Belshe RB (ed.) Textbook of human virology, 2nd Ed. Mosby Year Book, St. Louis, MO, pp. 791–803.

Liu OC Seraichekas HR, Akin EW, Brashear DA, Katz EL, Hill WJ (1971) Relative resistance of 20 human enteric viruses to free chlorine in Potomac River. In: Viruses and water quality: occurrence and control. Proceedings of the 13th water quality conference. University of Illinois Bulletin, No. 69, pp. 171–195.

Mackey JK, Green M, Wold WS, Rigden P (1979) Analysis of human cancer DNA for DNA sequences of human adenovirus type 4. J Natl Cancer Inst 62:23–26.

Martone WJ, Hierholzer JC, Keenlyside RA, Fraser WD, D'Angelo LJ, Winkler WG (1980) An outbreak of adenovirus type 3 disease at a private recreation center swimming pool. Am J Epidemiol 111:229–237.

McMinn PC, Stewart J, Burrell CJ (1991) A community outbreak of epidemic keratoconjunctivitis in central Australia due to adenovirus type 8. J Infect Dis 164:1113–1118.

McNeill KM, Hendrix RM, Lindner JL, Benton FR, Monteith SC, Tuchscherer MA, Gray GC, Gaydos JC (1999) Large, persistent epidemic of adenovirus type 4-associated respiratory disease in U.S. army trainees. Emerg Infect Dis 5:798–801.

Meiklejohn G (1983) Viral respiratory disease at Lowry Air Force Base in Denver, 1952–1982. J Infect Dis 148:775–784.

Meng QS, Gerba CP (1996) Comparative inactivation of enteric adenoviruses, polioviruses and coliphages by ultraviolet irradiation. Wat Res 30:2665–2668.

Michaels MG, Green M, Wald ER, Starzl TE (1992) Adenovirus infection in pediatric liver transplant recipients. J Infect Dis 165:170–174.

Mistchenko AS, Huberman KH, Gomez JA, Grinstein S (1992) Epidemiology of enteric adenovirus infection in prospectively monitored Argentine families. Epidemiol Infect 109:539–546.

Mizuta K, Suzuki H, Ina Y, Yazaki N, Sakamoto M, Katsushima N, Numazaki Y (1994) Six-year longitudinal analysis of adenovirus type 3 genome types isolated in Yamagata, Japan. J Med Virol 42:198–202.

Munoz FM, Piedra PA, Demmler GJ (1998) Disseminated adenovirus disease in immunocompromised and immunocompetent children. Clin Infect Dis 27:1194–2000.

Murtagh P, Cerqueiro C, Halac A, Avila M, Kajon A (1993) Adenovirus type 7h respiratory infections: a report of 29 cases of acute lower respiratory disease. Acta Paediatr 82:557–561.

Myrmel M, Berg EMM, Grinde B, Rimstad E (2006) Enteric viruses in inlet and outlet samples from sewage treatment plants. J Water Health 4:197–209.

National Research Council (1999) Setting Priorities for Drinking Water Contaminants. National Academy Press, Washington, DC.

Numazaki Y, Kumasaka T, Yano N, Yamanaka M, Miyazawa T, Takai S, Ishida N (1973) Further study of acute hemorrhagic cystitis due to adenovirus type 11. N Engl J Med 289:344–347.

Papapetropoulou M, Vantarakis AC (1998) Detection of adenovirus outbreak at a municipal swimming pool by nested PCR amplification. J Infect 36:101–103.

Payne SB, Grilli EA, Smith AG, Hoskins TW (1984) Investigation of an outbreak of adenovirus type 3 infection in a boys' boarding school. J Hyg (Lond) 93:277–283.

Pham TT, Burchette JL, Jr, Hale LP (2003) Fatal disseminated adenovirus infections in immunocompromised patients. Amer J Clin Pathol 140:575–583.

Pierdnoir E, Bureau-Chalot F, Merle C, Gotzamanis A, Wuibout J, Bajolet O (2002) Direct costs associated with a nosocomial outbreak of adenoviral conjunctivitis infection in a long-term care institution. Am J Infect Control 30:407–410.

Pina S, Puig M, Lucena F, Jofre J, Girones R (1998) Viral pollution in the environment and in shellfish: human adenovirus detection by PCR as an index of human viruses. Appl Environ Microbiol 64:3376–3382.

Pring-Åkerblom P, Adrian T (1995) Sequence characterization of the adenovirus 31 fibre and comparison with serotypes of subgenera A to F. Res Virol 146:343–354.

Puig M, Jofre J, Lucena F, Allard A, Wadell G, Girones R (1994) Detection of adenoviruses and enteroviruses in polluted waters by nested PCR amplification. Appl Environ Microbiol 60:2963–2970.

Pusch D, Oh DY, Wolf S, Dumke R, Schroter-Bobsin U, Hohne M, Roske I, Schreier E (2005) Detection of enteric viruses and bacterial indicators in German environmental waters. Arch Virol 150:929–947.

Rahal JJ, Millian SJ, Noriega ER (1976) Coxsackievirus and adenovirus infection. Association with acute febrile and juvenile rheumatoid arthritis. JAMA 235:2496–2501.

Ray CG, Holberg CJ, Minnich LL, Shehab ZM, Wright AL, Taussig LM (1993) Acute lower respiratory illnesses during the first three years of life: potential roles for various etiologic agents. The group medical associates. Pediatr Infect Dis 12:10–14.

Regli S, Rose JB, Haas CN, Gerba CP (1991) Modeling the risk from Giardia and viruses in drinking water. J Am Water Works Assoc 83:76–84.

Richmond SJ, Caul EO, Dunn SM, Ashley CR, Clark SK, Seymour NR (1979) An outbreak of gastroenteritis in young children caused by adenoviruses. Lancet 1:1178–1181.

Rodriguez WJ, Kim HW, Brandt CD, Schwartz RH, Gardner MK, Jefferies B, Parrott RH, Kaslow RA, Smith JI, Takiff H (1985) Fecal adenovirus from a longitudinal study of families in metropolitan Washington, DC: laboratory, clinical, and epidemiological observations. J Pediatr 107:514–520.

Roessler PF, Severin BF (1996) Ultraviolet light disinfection of water and wastewater. In: Hurst CJ (ed.) Modeling disease transmission and its prevention by disinfection. Cambridge University Press, Cambridge, pp. 313–368.

Rowe WP, Huebner RJ, Gillmore LK (1953) Isolation of a cytopathogenic agent from human adenoids undergoing spontaneous degeneration in tissue culture. Proc Soc Exp Biol Med 84:570–573.

Ruuska T, Vesikari T (1991) A prospective study of acute diarrhoea in Finnish children from birth to 2 ½ years of age. Acta Paediatr Scand 80:500–507.

Saad RS, Demetris AJ, Kusne S, Randhawa PS (1997) Adenovirus hepatitis in the adult allograft liver. Transplantation 64:1483–1485.

Schmitz H, Wigand R, Heinrich W (1983) Worldwide epidemiology of human adenovirus infections. Am J Epidemiol 117:455–466.

Scott-Taylor TH, Hammond GW (1995) Local succession of adenovirus strains in pediatric gastroenteritis. J Med Virol 45:331–338.

Sedmak G, Bina D, Macdonald J, Coulillard L (2005) Nine-year study of the occurrence of culturable viruses in source water for two drinking water treatment plants and the influent and effluent of a wastewater treatment plant in Milwaukee, Wisconsin (August 1994 through July 2003). Appl Environ Microbiol 71:1042–1050.

Shenk T (2001) Adenoviridae: the viruses and their replication. In: Knipe DM, Howley PM, Griffin DE, Lamb RA, Martin MA, Roizman B, Straus SE (eds.) Fields Virology, 4th Ed. Lippincott Williams & Wilkins, Philadelphia, PA, pp. 2265–2300.

Shields AF, Hackman RC, Fife KH, Corey L, Meyers JD (1985) Adenovirus infections in patients undergoing bone-marrow transplantation. N Engl J Med 312:529–533.

Shinozaki T, Araki K, Fujita Y, Kobayashi M, Tajima T, Abe T (1991) Epidemiology of enteric adenoviruses 40 and 41 in acute gastroenteritis in infants and young children in the Tokyo area. Scand J Infect Dis 23:543–547.

Singh-Naz N, Rodriguez W (1996) Adenoviral infections in children. Adv Ped Infect Dis 11:365–388.

Smith JC, Haddix AC, Teutsch SM, Glass RI (1995) Cost effective analysis of a rotavirus immunization program for the United States. Pediatrics 96:609–615.

Tani N, Dohi Y, Kurumatani N, Yonemasu K (1995) Seasonal distribution of adenoviruses, enteroviruses and reoviruses in urban river water. Microbiol Immunol 39:577–580.

Thurston-Enriquez JA, Haas CN, Jacangelo J, Gerba CP (2003a) Chlorine inactivation of adenovirus type 40 and feline calicivirus. Appl Environ Microbiol 69:3979–3985.

Thurston-Enriquez JA, Haas CN, Jacangelo J, Riley K, Gerba CP (2003b) Inactivation of feline calicivirus and adenovirus type 40 by UV radiation. Appl Environ Microbiol 69:577–582.

Thurston-Enriquez JA, Haas CN, Jacangelo J, Gerba CP (2005a) Inactivation of enteric adenovirus and feline calicivirus by chlorine dioxide. Appl Environ Microbiol 71:3100–3105.

Thurston-Enriquez JA, Haas CN, Jacangelo J, Gerba CP (2005b) Inactivation of enteric adenovirus and feline calicivirus by ozone. Water Res 39:3650–3656.

Turner M, Istre GR, Beauchamp H, Baum M, Arnold S (1987) Community outbreak of adenovirus type 7a infections associated with a swimming pool. South Med J 80:712–715.

Uhnoo I, Wadell G, Svensson L, Johansson ME (1984) Importance of enteric adenoviruses 40 and 41 in acute gastroenteritis in infants and young children. J Clin Microbiol 20:365–372.

Van R, Wun CC, O'Ryan ML, Matson DO, Jackson L, Pickering LK (1992) Outbreaks of human enteric adenovirus types 40 and 41 in Houston day care centers. J Pediatr 120:516–521.

Vangipuram SD, Sheele J, Atkinson RL, Holland TC, Dhurandhar NV (2004) A human adenovirus enhances preadiocytes differentiation. Obes Res 12:770–777.

van Heerden J, Ehlers MM, van Zyl WV, Grabow WOK (2003) Incidence of adenoviruses in raw and treated water. Water Res 37:3704–3708.

van Heerden J, Ehlers MM, Grabow WOK (2005a) Detection and risk assessment of adenoviruses in swimming pool water. J Appl Microbiol 99:1256–1264.

van Heerden J, Ehlers MM, Hiem A, Grabow WOK (2005b) Prevalence, quantification and typing of adenoviruses detected in river and treated drinking water in South Africa. J Appl Microbiol 99:234–242.

van Heerden J, Ehlers MM, Vivier JC, Grabow WOK (2005c) Risk assessment of adenovirus detected in treated drinking water and recreational water. J Appl Microbiol 99:926–933.

van Regenmortel MHV, Fauquet CM, Bishop DHL, Carstens EB, Estes MK, Lemon SM, Maniloff J, Mayo MA, McGeoch DJ, Pringle CR, Wickner RB (2000) Virus Taxonomy: Seventh Report of the International Committee on Taxonomy of Viruses. Academic Press, San Diego, CA, pp. 227–238.

Villena C, Gabrieli R, Pinot RM, Guix S, Donia D, Buonomo E, Palombi L, Cenko F, Bino S, Bosch A, Divizia M (2003) A large infantile gastroenteritis outbreak in Albania caused by multiple emerging rotavirus genotypes. Epidemiol Infect 131:1105–1110.

Wadell G (1994) Molecular epidemiology of human adenoviruses. Curr Top Microbiol Immunol 110:191–220.

Wang B, Chen X (1997) The molecular epidemiological study on enteric adenovirus in stool specimens collected from Wuhan area by using digoxigen in labeled DNA probes. J Tongji Med Univ 17:79–82.

Wang Y, Krushel LA, Edelman GM (2003) Targeted DNA recombination in vivo using an adenovirus carrying the cre recombinase gene. Proc Natl Acad Sci USA 93:3932–3936.

Webb DH, Shields AF, Fife KH (1987) Genomic variation of adenovirus type 5 isolates recovered from bone marrow transplant recipients. J Clin Microbiol 25:305–308.
Wold WS, Mackey JK, Rigden P, Green M (1979) Analysis of human cancer DNA's for DNA sequence of human adenovirus serotypes 3, 7, 11, 14, 16, and 21 in group B1. Cancer Res 39:3479–3484.
Yates MV, Rochelle P, Hoffman R (2006) Impact of adenovirus ressitance on UV Disinfection requirements: a report on the state of adenovirus science. J Amer Water Works Assoc 98:93–106.
Zahradnik JM, Spencer MJ, Porter DD (1980) Adenovirus infection in the immuno-compromised patient. Am J Med 68:725–732.

Haloacetonitriles: Metabolism and Toxicity

John C. Lipscomb,(✉) Ebtehal El-Demerdash, and Ahmed E. Ahmed

Contents

1 Introduction ... 169
2 Formation and Prevalence of HANs in Drinking Water ... 170
  2.1 Formation ... 170
  2.2 Prevalence in Drinking Water ... 171
3 Pharmacokinetics and Metabolism ... 171
  3.1 Pharmacokinetics ... 171
  3.2 Metabolism ... 174
4 General and Systemic Toxicity of HANs ... 176
  4.1 Acute Toxicity ... 176
  4.2 Reproductive and/or Developmental Effects ... 178
  4.3 HANs-Induced Gastrointestinal Tract Injury ... 181
5 Genotoxicity and Carcinogenesis of HANs ... 183
  5.1 Genotoxic Effects ... 183
  5.2 Carcinogenic Effects ... 186
6 Mechanism of Action ... 188
  6.1 General Mechanistic Pathways ... 188
  6.2 Oxidative Stress as a Signaling Pathway for HANs-Induced Adverse Effects ... 190
7 Comparison of the Relative Chemical and Biological Activities of HANs ... 192
8 Summary ... 194
References ... 195

1 Introduction

Chlorination to purify water supplies is among the most important public health advances of the twentieth century. Following the introduction of widespread water chlorination, in 1908, once-common diseases such as cholera, dysentery and typhoid fever were practically eliminated (Mughal 1992). However, formation of disinfection by-products (DBPs), which result from the interaction of chlorine with organic materials in the source water, is a side effect and major hazard of chlorination

U.S. Environmental Protection Agency, National Center for Environmental Assessment
Cincinnati, Ohio 45268, USA

D.M. Whitacre (ed.) *Reviews of Environmental Contamination and Toxicology.*

(Doyle et al. 1997). Recent epidemiological studies provide moderate evidence that DBPs cause adverse pregnancy outcomes (APO) (Bove et al. 2002). Results of experimental studies indicate that some DBPs are mutagens, carcinogens, teratogens, or developmental toxicants (Ahmed et al. 2005a, b; Muller-Pillet et al. 2000; Villanueva et al. 2004). The Safe Drinking Water Act authorized the U.S. Environmental Protection Agency (U.S. EPA) to enforce drinking water regulations. In 2006, the Stage 2 Disinfectants/DBPs Rule was promulgated (U.S. EPA 2006). To comply with this rule, some utilities are switching from chlorine to chloramine disinfection, which may increase nitrogenous DBPs (N-DBPs). N-DBPs were cited as research priorities by the U.S. EPA (Weinberg et al. 2002; Woo et al. 2002).

The purpose of this review is to advance the goal of defining human health risks of these chemicals. We present and focus on one class of N-DBPs, the halogenated acetonitriles (HANs). We address the following information on HANs: formation and prevalence in drinking water, general and systemic toxicity, mutagenicity, carcinogenicity, pharmacokinetics and metabolism, mechanism of action and evaluation of other toxic effects. Moreover, in this review we present a comparative analysis of the biological activity of individual HANs.

2 Formation and Prevalence of HANs in Drinking Water

2.1 Formation

The chlorination of water commonly results in the formation of several HANs (X_nCH_nCN): monobromoacetonitrile ($BrCH_2CN$; BAN), bromochloroacetonitrile (BrClCHCN; BCAN), dibromoacetonitrile (Br_2CHCN; DBAN), monochloroacetonitrile ($ClCH_2CN$; CAN), dichloroacetonitrile (Cl_2CHCN; DCAN), and trichloroacetonitrile (Cl_3CCN; TCAN) (Oliver 1983; Trehy and Bieber 1981). Factors that influence HAN formation during the chlorination process include: temperature, pH, time, chlorine and bromide ion concentrations, and content of organic and inorganic materials in source water (Yang et al. 2007). The presence of bromide ions in aqueous solutions of chlorine results in preferential substitution reactions with bromide and organic substrates present in water; such reactions lead to the formation of various bromine-containing HANs (Coleman et al. 1984; Pourmoghaddas et al. 1993). Moreover, residual chlorine (hypochlorite) in drinking water may result in the formation of DBAN and DCAN *in vivo*, following human consumption of such water (Mink et al. 1983). HANs degrade by gradual hydrolysis to form nonvolatile products; degradation rates increase with increased water pH (Bieber and Trehy 1983).

HANs are not known to occur naturally. Synthetic HANs have limited industrial and agricultural use in pesticide fumigation (Pereira et al. 1984; Sax and Lewis 1987; Trehy et al. 1986). In addition, DCAN has been used as an insecticide for grains (Cotton and Walkden 1968) and as a biological growth inhibitor in cooling towers (Matt 1968).

2.2 *Prevalence in Drinking Water*

Knowing the concentration of HANs in drinking water supplies is essential in defining levels and routes of human exposure and ultimately the extent of their adverse effects on human health. However, it is challenging to accurately predict exposure to DBPs, because their occurrence and concentrations are highly dependent on the disinfection method used. In addition, other variables exist: amount of organic matter in the water, temperature of the water, and distance over which the water is piped after disinfection to the site of use (Bove et al. 2002). Furthermore, human exposure to HANs is not limited to oral consumption of drinking water; dermal and inhalation exposures also occur while showering, swimming and other household activities (Bruchet et al. 1985).

HANs are routinely detected in drinking water supplies in the U.S. and internationally (McKinney et al. 1976; Otson 1987; Reding et al. 1989; Stevens et al. 1990; Suffet et al. 1980; Trehy and Bieber 1981; Trehy et al. 1986, 1987; Weinberg et al. 2002). CAN, BCAN, DBAN and TCAN are the HAN analogs most commonly found, and they have been detected in several surveys: a U.S. survey of 35 water utilities (Krasner et al. 1989), a survey of 53 Canadian water utilities (Williams et al. 1997) and the U.S. EPA's Information Collection Rule effort, which involved 500 large drinking water plants in the U. S. (McGuire et al. 2002). These four above- mentioned HANs were formed at plants that used chlorine, chloramine, chlorine dioxide or ozone for disinfection. Plants using chloramines (with and without chlorine) had the highest levels of HANs in their finished drinking water. The relative concentrations of individual HANs varied considerably among drinking water supplies (Table 1), and levels in water supplies that had been lime-softened were not detectable.

HANs were also detected in other water reservoirs such as swimming pools (Weisel et al. 1999). A recent study indicated that the formation of DCAN in swimming pools resulted from increased chlorination-mediated reactions that degraded nitrogen-containing compounds of human origin; these compounds included urea, proteins (hair), lotion, saliva and skin (Kim et al. 2002). Residues of DBAN, BAN and BCAN were also detected in foods and beverages. To facilitate the investigation of human exposure to DBPs via foods and beverages, analytical recovery methods for HANs and other DBPs were developed. HANs were generally well recovered (70–130%) from food products spiked with authentic HANs standards, with the exception of BCAN (64%) and DBAN (55%) (Raymer et al. 2000).

3 Pharmacokinetics and Metabolism

3.1 *Pharmacokinetics*

To evaluate the bioavailability in target organs of various HANs, absorption and distribution studies were conducted following oral or intravenous (iv) administration of radiolabeled HANs molecules (Ahmed et al. 1991b, 1992; Abdel-Aziz et al. 1993; Lin et al. 1992; Roby et al. 1986).

Table 1 Occurrence of halogenated acetonitrile (HAN) compounds in drinking water[a]

	Haloacetonitrile				
Treatment	BCAN	DBAN	DCAN	TCAN	HAN4[b]
Overall/All[c]	(1,294; 407)[d]	(1,288; 407)	(1,285; 409)	(1,279; 409)	(1,257; 407)
Mean	1.02[e]	0.78	1.89	0.08	3.73
Median	0.78	0.28	1.30	0.00	3.0
SD (standard deviation)	1.17	1.16	2.28	1.33	3.56
90th percentile	2.43	2.33	4.40	0.00	7.85
SW/All[f]	(927; 297)[d]	(922; 297)	(918; 297)	(918; 297)	(901; 297)
Mean	1.11	0.74	2.26	0.03	4.08
Median	0.88	0.24	1.73	0.00	3.33
SD	1.17	1.13	2.20	0.26	3.13
90th percentile	2.45	2.35	4.63	0.00	8.0
GW/All[g]	(350; 108)[d]	(349; 108)	(350; 110)	(334; 110)	(339; 108)
Mean	0.76	0.83	0.93	0.21	2.75
Median	0.06	0.40	0.00	0.00	0.85
SD	1.13	1.24	2.25	2.53	4.42
90th percentile	2.14	2.18	2.83	0.00	7.39

[a] Data adapted from the U.S. EPA, Office of Groundwater and Drinking Water, Information Collection Rule (ICR) Data Analysis Technical Working Group. This database contains the quarterly Distribution System average, which is an average of the four physical locations sampled in the distribution systems that connect each of the municipal drinking water plants with their community end-users. Samples were taken at points representing the average, and the maximal distances from the plants. Minimum reporting levels for each of the halogenated acetonitriles (HANs) was 0.5 μg /L. Analysis for the HANs was accomplished using U.S. EPA method 551.1 (U.S. EPA 1995)

[b] Data are presented as ppb for the combined four HANs quantified

[c] Data represent all treatment systems, irrespective of source water or treatment

[d] Numbers in parentheses represent the number of quarterly samples taken; number of plants submitting samples

[e] Data are presented as ppb

[f] *SW* Surface water. Data represent all treatment systems that use a surface water source, irrespective of treatment

[g] *GW* Ground water. Data represent all treatment systems that use a groundwater source, irrespective of treatment

Roby and colleagues (1986) studied the excretion and tissue distribution of DCAN administered orally, in water, to male F344 rats and male B6C3F1 mice. Three dose levels (0.2, 2.0, and 15 mg/kg) were employed with rats, and two dose levels with mice (2.0 and 15 mg/kg) with both [1-^{14}C]- and [2-^{14}C]-labeled DCAN. [1-^{14}C]-DCAN is labeled at the cyanide (CN) carbon, and [2-^{14}C]-DCAN is labeled at the dichloromethyl carbon. All excreta were collected until >70% of the dose had been recovered. Animals were then sacrificed, and the distribution of radioactivity among body organs determined at various time intervals. DCAN was rapidly absorbed after oral administration in a water vehicle. Differences in elimination between the two labeling positions were observed for both rats and mice. In rats, the C-2 (dichloromethyl) DCAN label was eliminated appreciably faster. Liver tissues retained the highest levels of ^{14}C, with the exception of blood from rats dosed with [1-^{14}C]-DCAN. Differences in the route of excretion between the [1-^{14}C]- and [2-^{14}C]-DCAN indicated that the molecule was cleaved in the body and was subsequently metabolized by different mechanisms (Roby et al. 1986). Results also indicated that DCAN is highly metabolized in the liver to various excretable forms. The higher accumulation of radioactivity in liver, skin and muscle suggests that these organs are target sites for DCAN accumulation (bioavailability) and HANs toxicity.

Lin et al. (1992) administered oral doses of 1-or 2-[^{14} C]-TCAN in tricaprylin to male Fischer 344 rats. To determine the nature of the interactions between HANs andmacromolecules, several protein fractions (globin, albumin and globulins) and DNA were isolated from blood and solid tissues of treated animals. [^{14} C]-TCAN or its metabolites were found to bind to both DNA and blood proteins in a dose-related manner. Higher levels of ^{14}C activity (DPM [disintegrations per minute]/μg DNA) were associated with the DNA of liver, kidney and stomach, when the carbon was labeled at the C_2 position rather than at the C_1 position. The stomach exhibited the highest level of DNA binding, followed by the liver and kidney. However, the position of the radiolabel did not influence levels of radioactivity associated with blood proteins. In addition, binding of ^{14}C-TCAN equivalents was higher in DNA isolated from rats killed 24 hr after dosing than those killed 4 hr after administration. In contrast, the three blood proteins (globin, albumin, and globulins) showed similar binding levels, regardless of exposure time. Radioactivity associated with DNA was not incorporated into the nitrogen bases, i.e., as a result of de novo synthesis (Lin et al. 1992).

The metabolic fate of 2-[^{14}C[-CAN was investigated in mice following iv administration; 12 hr after administration, 0.8% of the dose was exhaled as unchanged CAN, and 65.7% of administered radioactivity was excreted in urine, feces and exhaled as $^{14}CO_2$. ^{14}C-CAN was not detected in urine or feces. In addition, extensive uptake and retention of CAN equivalents were observed in several organs. The retention of radioactivity in tissues of the thyroid gland, gastrointestinal tract (GIT), testes, brain and eye suggest that those organs are potential targets for CAN bioavailability and target organ toxicity (Ahmed et al. 1992; Bhat et al. 1990).

To compare the fate and bioavailability of CAN with that of its major metabolites and carbon dioxide, the metabolic fate of CAN and Na_2CO_3 were investigated (Ahmed et al. 1991b) using whole body autoradiographic techniques (Ullberg 1977). Five min after CAN administration, extensive accumulation of radioactivity occurred in liver, kidney and the GIT. After 3 hr, radioactivity had diffused homogeneously to all tissues. At 24 and 48 hr, radioactivity was refocused in liver, kidney, GIT walls, bone marrow, eye and thyroid gland. It is clear from these studies that CAN undergoes extensive metabolic biotransformation in rats.

Abdel-Aziz et al. (1993) employed diethylmaleate-induced depletion of reduced glutathione (GSH) to examine the affect of GSH on tissue binding and disposition of ^{14}C-CAN in mice. Depletion of GSH resulted in a fivefold higher elimination in urine of the CAN metabolite, thiocyanate (SCN), when assessed at 24, 48 and 72 hr; this result indicates apparent competition between GSH conjugation and the metabolic oxidative path leading to SCN formation, and is consonant with research by Pereira et al. (1984), who suggest that HANs are metabolized via the mixed function oxidase system.

3.2 Metabolism

Metabolism of HANs to CN, and release of CN^- from organic nitriles leading to the formation of aldehydes, has been proposed by several authors (International Agency for Research on Cancer; IARC 1991). The released CN^- is further metabolized by rhodanases to SCN. When single doses of various HANs (0.75 mmole/kg) were administered orally to Sprague-Dawley rats, the relative extent of SCN excretion was CAN > BCAN > DCAN > DBAN > TCAN (Lin et al. 1986). The proposed pathway for CN^- involved formation of hydroxyacetonitriles, either by displacement of a halide ion by a hydroxyl group or by oxidation of a hydrogen atom through the action of a mixed-function oxidase (Fig. 1). The metabolism of these compounds to SCN results in release of highly reactive metabolites, which may, themselves, be responsible for carcinogenic or other toxic effects (Lin and Guian 1989; Silver et al. 1982; Swenberg et al. 1980). It is proposed that monohaloacetonitrile yields formaldehyde, dihaloacetonitrile yields formyl CN or formyl halide, and TCAN yields phosgene or cyanoformyl chloride. In addition, mono- and DCANs may be metabolized by mixed-function oxidases to yield cyanochloromethanol (ClCH(OH)CN) and cyanodichloromethanol (Cl_2C-(OH)CN), respectively. The proposed intermediates are direct-acting alkylating agents, are highly toxic, and at least in the case of formaldehyde, is a known carcinogen (Silver et al. 1982) There is a discrepancy between the alkylation potential and extent of metabolism of CAN and TCAN to SCN; the discrepancy may result from the much greater chemical reactivity of phosgene (Cl_2CO), a proposed TCAN metabolite, and cyanoformyl chloride (ClCO-CN), compared to another proposed CAN metabolite, formaldehyde (Pereira et al. 1984). Therefore, the relative biological activity of HANs may well be defined by the chemical reactivity of both the parent chemicals and their reactive metabolic intermediates.

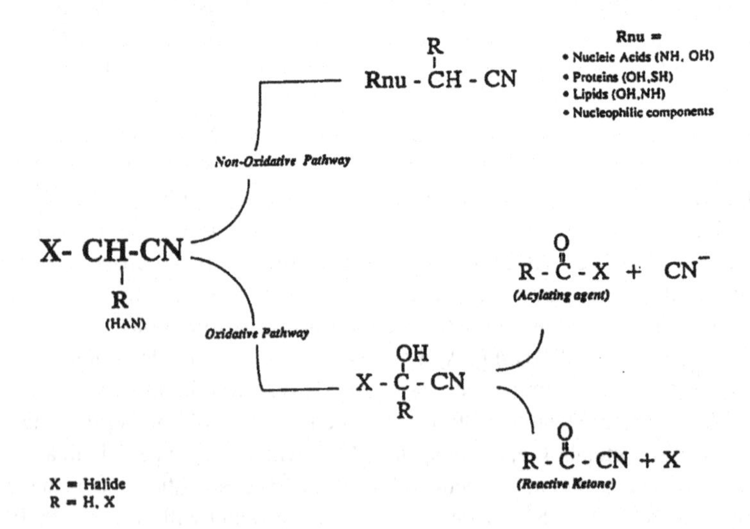

Fig. 1 Metabolic pathways for halogenated acetonitriles (HANs)

Pereira et al. (1984) examined the interactions of DBAN, BCAN, DCAN, TCAN and CAN with the cytochrome P-450 enzyme family; the key measure was the affect of these compounds on dimethylnitrosamine demethylase (DMN-DM) activity, a marker selective for cytochrome P-450 2E1 (CYP2E1). An investigation of the potential of these compounds to inhibit DMN-DM revealed that DBAN and BCAN were the most potent inhibitors, with I_{50} s (concentration that inhibits 50% of the enzymatic activity) of 3–4 × 10^{-5} M (Pereira et al. 1984). The I_{50} value for DCAN and TCAN was more than tenfold higher; CAN inhibited DMN-DM activity the least. Because of the apparent high reactivity of HANs, substantive accumulation of these compounds in intact liver seems unlikely. The I_{50} values, as determined *in vitro,* may represent gross quantitative overestimates of the concentrations necessary to produce 50% inhibition *in vivo*. To confirm the potential of HANs to produce a similar effect *in vivo*, rats were dosed with 0.75 mmol/kg of TCAN and DBAN, and enzymatic fractions (microsomes) were isolated from the livers at 3 and 18 hr post-dosing. DMN-DM activity was significantly decreased at both time intervals following TCAN administration, and was significantly increased 18 hr after DBAN administration. Based on their findings, Pereira et al. (1984) have proposed a metabolic scheme for the formation of SCN, which involves an initial hydroxylation of HANs by the hepatic mixed-function oxidase system.

The effect of HANs on glutathione-S-transferases (GSTs) has been studied *in vitro* and *in vivo* (Ahmed et al. 1989, 1991a; Lin and Guian 1989). The substrate 1-chloro-2,4-dinitrobenzene (CDNB) is conjugated with GSH by many GST isoforms. Although CAN and BAN were poor inhibitors of GST activity toward CDNB, DCAN, TCAN and DBAN were good inhibitors of GST *in vitro*, with I_{50} concentrations of 2.5 × 10^{-3}, 3.4 × 10^{-4}, and 8.2 × 10^{-4} M, respectively (Ahmed et al. 1989). Further investigation determined that the activity was completely restored when the enzyme was dialyzed. These results suggest that the degree of inhibition of this enzyme pathway is dependent on HAN concentrations in the liver at a given time, and inhibition ceases following the elimination of HANs. Because CDNB is conjugated with GSH by several GST isoforms, data describing inhibition of GST-mediated CDNB conjugation activity cannot identify a GST class- or isoform-specific interaction. The apparent reversibility of GST inhibition contrasts with the irreversible nature of the effect of HANs on the CYP2E1-mediated metabolism of DMN, as demonstrated by Pereira et al. (1984).

Another HAN metabolism pathway was discovered and is mediated by oxidative bioactivation of HAN molecules (Lin et al. 1986). To verify this pathway, Mohamadin (2001) investigated the role of reactive oxygen species (ROS) in activation of DCAN. A model ROS generation system (Fenton-like reaction; Fe_2^+ and H_2O_2), that mainly produces hydroxyl radicals, was used. DCAN oxidation was monitored by measuring the extent of CN^- release. Results indicate that DCAN was markedly oxidized by this system, and the rate of oxidation was dependent on the concentration of DCAN. A fourfold increase in H_2O_2 concentration (50–200 mM) resulted in a 35-fold increase in CN^- release. The rates of DACN oxidation, in the presence of various transition metals, were in the following order: iron > copper > titanium. DCAN oxidation was enhanced significantly by the addition of Vitamin C and sulfhydryl compounds. Conversely, the addition of an H_2O_2 scavenger, iron chelator or antioxidants resulted in a significant decrease in CN^- release. Michaelis-Menten kinetic analysis of reaction rates, with or without inhibitors, indicated that ROS-mediated oxidation of DCAN was inhibited by catalase ($K_i = 0.01$ mM) > desferroxamine (0.02 mM) > mannitol (0.09 mM) > dimethylsulfoxide (0.12 mM). These results indicate that DCAN is oxidized by a ROS-mediated mechanism that may be important in DCAN bioactivation. Because cellular stress (i.e., oxidative stress) may reduce the effectiveness of repair processes, it is possible that this situation may exacerbate DCAN-induced genotoxicity at target organs, where ROS generating systems are abundant.

4 General and Systemic Toxicity of HANs

4.1 Acute Toxicity

Acute toxicity data on HANs (Ahmed et al. 1991a; Hayes et al. 1986; Meier et al. 1985a; Simmon et al. 1977) suggested that they may be biologically reactive, either directly or following enzymatic activation.

Toxicity studies by Tanii and Hashimoto (1984) indicated that ataxia, depressed respiration, decreased activity, and coma are symptoms that precede death after animals are intoxicated by HANs. The LD_{50} values of different HANs are shown in Table 2. HANs LD_{50} values were affected by the dosing vehicle used. Moreover,

Table 2 LD_{50} values for HANs in rats or mice

Compound	Species	LD_{50} (mg/kg BW)	Route/vehicle	Reference
CAN	Mice	136	Oral	Suffet et al. 1980
DBAN	Rats (M)	245	Oral/CO	Hayes et al. 1986
	Rats (F)	361	Oral/CO	Hayes et al. 1986
	Mice (M)	289	Oral/CO	Hayes et al. 1986
	Mice (F)	303	Oral/CO	Hayes et al. 1986
DCAN	Rats (M)	339	Oral/CO	Hayes et al. 1986
	Rats (F)	330	Oral/CO	Hayes et al. 1986
	Mice (M)	270	Oral/CO	Hayes et al. 1986
	Mice (F)	279	Oral/CO	Hayes et al. 1986
BAN	Rats (M)	26	Oral/DMSO	Ahmed and Hussein 1987
DBAN	Rats (M)	99	Oral/DMSO	Ahmed and Hussein 1987
DCAN	Rats (M)	202	Oral/DMSO	Ahmed and Hussein 1987
CAN	Rats (M)	153	Oral/DMSO	Ahmed and Hussein 1987

HANs halogenated acetonitriles, *BW* body weight, *M* Male, *CO* Corn oil, *F* Female, *DMSO* Dimethyl sulfoxide rats and mice are relatively sensitive to the acute effects of DCAN; however, male rats appear to be more sensitive to DBAN than female rats. When DBAN was administered by gavage to male and female CD rats at a dose of 90 mg/kg/d, there was 40% mortality in males and 20% in females. DBAN-induced mortality was 10% for males and 5% for females at a dose of 45 mg/kg. However, at 23 mg/kg/d, 10% of female rats died and no lethality was observed in males during the 90-d study (Hayes et al. 1986).

When the oral toxicity and lethality of various HANs were examined, there was an eightfold difference between LD_{50} values of the most and least toxic HAN analogs. The LD_{50} values (mg/kg BW) for these compounds were: BAN (26), DBAN (99), CAN (153) and DCAN (202) (Ahmed and Hussein 1987; Ahmed et al. 1991a). Signs of toxicity included gasping, salivation, chromodacryorrhea, increased urination and cyanosis. At higher doses, convulsions and death was observed within 48 hr after treatment. Histological examinations of tissues indicated severe necrosis in the submucosal layer of the gastro-esophageal junction. The authors suggested that the toxicity of HANs is a function of the type and number of halogen atom substitutions: dihalo-compounds were more toxic than monohalonitriles. Furthermore, bromo-compounds were more toxic than chloro derivatives (Ahmed and Hussein 1987; Ahmed et al. 1991a).

Nouraldeen et al. (1993) suggested that GIT tissues are potential target sites for the bioavailability, accumulation and toxicity of HANs. Histological examinations of animals treated with various doses of several HANs showed severe necrosis in the submucosal layer of the gastro-esophageal junction. Included in molecular toxicology studies were studies on the induction of unscheduled DNA synthesis (UDS), as a marker of genotoxicity, and the induction of replicative DNA synthesis (RDS) and ornithine decarboxylase activities (ODCs), as markers for induced cell proliferation in the glandular stomach mucosa of male rats following treatment with HANs. It was concluded that BAN, DBAN and CAN, at relatively high doses, produce epigenotoxic and cell proliferative effects in the rat glandular stomach mucosa.

The effect of different HANs on body weight, relative organ weights, clinical chemistry and hematological parameters was also evaluated. Repeated oral administration of DBAN or DCAN to rats revealed a dose-dependent decrease in weight gain. Because weight gain is a sensitive indicator for general animal health, the decrease suggests that the body responds negatively to the adverse effects of HANs. However, the mechanism of such effects is unclear (Hayes et al. 1986). In HAN-treated CD rats, an upward trend in values for hemoglobin, total red blood cell count, white blood cell count and fibrinogen was observed. However, results of the biochemical data failed to disclose the specific target organs or tissues. In addition, organ weight and organ weight/body weight ratio data suggest that the thymus, liver, spleen and gonads are possible target organs, whereas, histological data indicated that the GIT is a major target organ for HAN acute toxicity (Hayes et al. 1986).

4.2 Reproductive and/or Developmental Effects

Epidemiological studies indicate that exposure to DBPs in drinking water is associated with APO (Bove et al. 2002; Graves et al. 2001; King et al. 2000). The effects include congenital CNS anomalies such as neural tube defects (Klotz and Pyrch 1999); intra-uterine growth restriction (Graves et al. 2001; Kramer et al. 1992; Wright et al. 2004); skeletal defects (Magnus et al. 1999); and spontaneous abortion (Aschengrau et al. 1989). Significant association between APO and DBPs was found in women who consumed more than five glasses of water per d in which DBPs concentrations were ~75 μg/L (Waller et al. 1998; Wright et al. 2004). Therefore, to evaluate this causal relationship between maternal DBP exposures and APOs, it is essential to develop an animal model in which exposure to known concentrations of individual DBPs can be studied at various gestational stages.

A series of seven randomly selected DBPs were evaluated *in vitro* by the hydra assay to define their developmental toxicity potential (Fu et al. 1990). The authors predicted that six of the tested chemicals, including DBAN and TCAN, would be approximately equally toxic to both adult and embryonic mammals, when studied using standard developmental toxicity tests (teratology) (Fu et al. 1990).

In another series of investigations (George et al. 1985; Smith et al. 1986, 1987, 1988, 1989) HAN-induced developmental toxicity was examined in Long-Evans rats, and effects were compared with those of acetonitrile (ACN). The protocol for the *in vivo* teratology screen utilized (Chernoff and Kavlock 1982), involved exposing pregnant rats to a single maximum tolerated dose (MTD) between d 7 and 21 of gestation, and subsequent monitoring of reproductive success in dams and growth and viability in pups. The agents examined were ACN, CAN, BCAN, DBAN, DCAN and TCAN. Other than BCAN, all HANs induced maternal toxicity (reduction in maternal weight gain), when administered during gestation. ACN produced maternal toxicity only at the higher doses (300, 500 mg/kg), but did not induce any developmental effects. In addition, ACN, DCAN and TCAN all produced reductions in the number of viable litters delivered. However, no predictable relationship was observed between such reductions and the accompanying level of maternal toxicity. HANs caused reductions in birth weight and, in some instances, reduced weight gain over the first 4 d of postnatal life.

The results of the foregoing studies agreed with those of Smith et al. (1987), who evaluated the developmental toxicity of ACN and five of its halogenated derivatives using the same *in vivo* teratology screen. The screen was extended to evaluate pup growth until postnatal d 41–42 and weight of several organs, at sacrifice. Prenatal survival of the pups was adversely impacted by DCAN and TCAN. Postnatal growth, until d 4, was reduced by DCAN and BCAN; TCAN caused growth effects until d 42.

In serial studies, Smith et al. (1988, 1989) investigated the teratogenic effects of DCAN and TCAN in pregnant Long-Evans rats. Animals were dosed by oral intubation on gestation d 6–18, with doses ranging from 5–45 to 1–55 mg/kg/d for DCAN and TCAN, respectively. Tricaprylin was used as a dosing vehicle. The highest dose tested of DCAN and TCAN was lethal to 9 and 21% of dams, respectively.

DCAN caused resorption of the entire litter in 60% of surviving dams, whereas TCAN produced 100% resorptions in two-thirds of survivors. The incidence of soft tissue malformations was dose dependent. The observed anomalies were principally in the cardiovascular (interventricular septal defect, levocardia and abnormalities of the major vessels) and urogenital (hydronephrosis, rudimentary bladder and kidney, fused ureters, pelvic hernia and cryptorchidism) systems. The no-observed-adverse-effect level (NOAEL) for toxicity in pregnant Long-Evans rats was established by statistical analysis to be 15 mg/kg/d for DCAN and 1.0 mg/kg/d for TCAN (Smith et al.1988, 1989).

In 1996, Christ and coworkers evaluated the effect of vehicle administration on TCAN-induced developmental toxicity. These investigators reported that the TCAN dose-response curve for fetal (but not maternal) effects was shifted to the left (made more potent) when tricaprylin was used as vehicle instead of corn oil. Moreover, fetal weight was reduced when doses of 15 mg/kg TCAN dissolved in tricaprylin, was used; using corn oil as the vehicle, fetal weights were affected at does of about 55 mg/kg TCAN. When TCAN was administered in corn oil, the mean frequency of soft-tissue malformations decreased, when compared to TCAN administration in tricaprylin, and significantly fewer septal and great vessel cardiovascular defects were observed. It was proposed that TCAN may interact with the tricaprylin vehicle in a way that potentiates effects on the developing cardiovascular system. The NOAEL for TCAN (using corn oil as vehicle) was determined to be 35 kg/kg (Christ et al. 1996).

When comparing effects among HANs and to the parent compound (ACN), increasing halide substitution at the α-carbon appears to increase maternal and fetal toxicity (Christ et al. 1996; Moudgal et al. 2000; Smith et al. 1987, 1988). Similarities between type and extent of halogenation among HANs my help explain the rank-order of HANs potency, when using quantitative structure-toxicity programs to predict fetal toxicity or CN formation and SCN excretion. Increasing the chlorine substitution increased toxicity and decreased metabolism, potentially indicating an inverse correlation between metabolic activation and the developmental toxicity of these compounds (Lin et al. 1986).

Only limited mechanistic studies exist on the developmental toxicity of HANs. To understand the potential mechanisms involved in developmental toxicity of CAN, the disposition, transplacental uptake and irreversible interaction of CAN, and its metabolites, were evaluated in normal and GSH-depleted pregnant mice and fetuses (Abdel-Aziz et al. 1993; Jacob et al. 1998). Both normal and GSH-depleted (by administration of diethylmaleate) pregnant mice were given an iv dose of 2-[^{14}C]-CAN (333 μCi/kg equivalent to 77 mg/kg). Animals were processed for whole-body autoradiography (WBA) at 1, 8 and 24 hr after treatment. Tissue distribution of radioactivity was quantitated using computer-aided image analysis. A rapid and high uptake (at 1 hr) of radioactivity in normal and GSH depleted mice was observed in all major maternal tissues (liver, lung, urinary bladder, gastrointestinal mucosa, cerebellum, and uterine luminal fluid), and fetal tissues (liver, GIT and brain). This same pattern of distribution, though less intense, was observed at 8 hr following treatment. At 24 hr, there was a significantly higher retention and covalent interaction of radioactivity in GSH-depleted mouse tissues than in control tissues, particularly in the liver. The authors suggested that 2-[^{14}C]-CAN and/or its metabolites are capable of crossing the placental barrier. Enhancement of the molecular interaction and covalent binding of CAN in maternal and fetal DNA, following GSH depletion, indicates that GSH has an important role in CAN metabolism and contributes to subsequent adverse developmental and teratogenic effects.

Chemicals administered at low dose levels may cause molecular defects that disrupt developmental processes, without inducing observable physical malformations, though they lead to APO and/or teratogenic responses. In two recent studies, Ahmed et al. (2005a, b) examined the uptake and retention of [2-^{14}C]-CAN/metabolites (M) in fetal mouse brain and its developmental toxicity affects, using WBA. These authors also studied the effect of CAN administered to dams (gestation d 6–18) at daily low doses (25 mg/kg) that did not produce detectable maternal or fetal toxicity. WBA studies indicated significant uptake and retention of [2-^{14}C]-CAN/M in fetal brain (cerebral cortex, hippocampus, cerebellum) at 1 and 24 hr after dosing (Fig. 2). Compared with controls, fetuses exposed to CAN experienced a 20% reduction in body weight and a 22% reduction in brain weight. A significant increase in oxidative stress markers was observed in various fetal brain regions, as indicated by a three- to fourfold decrease in the ratio of GSH/GSSG (GSSG, oxidized glutathione), increased lipid peroxidation (1.3-fold), and

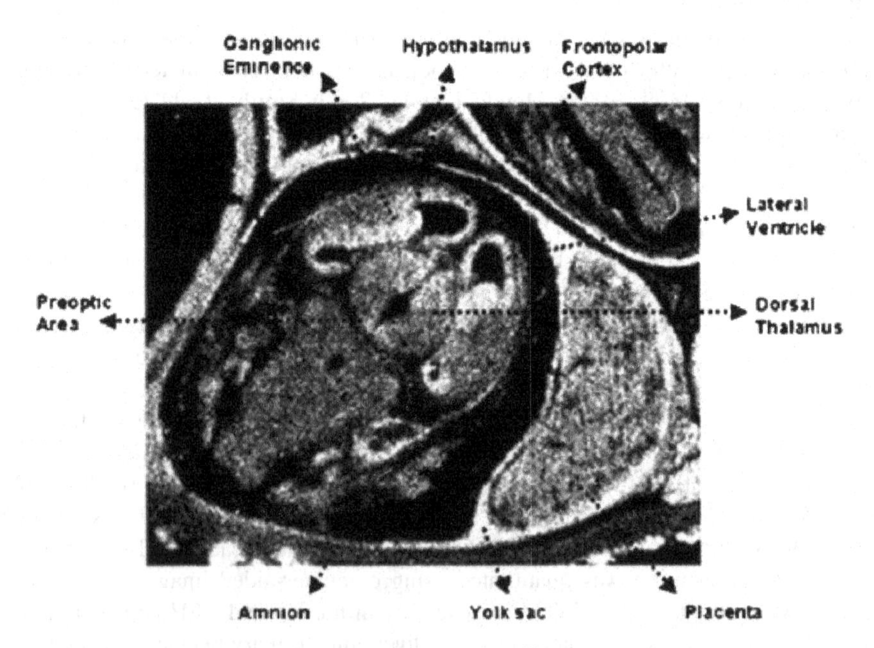

Fig. 2 In situ micro-whole-body autoradiograph of [2-^{14}C]-CAN 24 h after intravenous (iv) administration to a pregnant mouse at gestation d 12. High levels of [^{14}C]-equivalents (white areas) from [^{14}C]-CAN can be observed in frontal cortex, hypothalamus, ganglionic eminence, thalamus, preoptic area, yolk sac, and placenta

AU5,6

increased 8-hydroxy-2-deoxyguanosine levels (1.4-fold). In exposed animals, cupric silver staining results showed a significant increase in the number of degenerating neurons in cortical regions of the fetal brain. Moreover, there was an increase in nuclear DNA fragmentation (twofold increase in apoptotic indices) and caspase activity (1.5–1.7-fold) in fetal cerebral cortex and cerebellum. The authors concluded that CAN/M crossed the placenta and accumulated in fetal brain tissues, where it caused oxidative stress and neuronal apoptosis.

4.3 HANs-Induced Gastrointestinal Tract Injury

The physiology and immune function of GIT tissues make them unique in their expression of ROS-generating systems. The gut-associated lymphoid tissue (GALT) generates massive amounts of ROS, both spontaneously, and as a result of pathological events such as chemically-induced inflammation and neutrophilic activation (Grisham and Granger 1988). Other sources of ROS in the GIT are mucosal oxidases such as xanthine oxidase and aldehyde oxidase (Krenitsky et al. 1974). Generation of free radicals has been implicated in the activation of ACN, and other nitriles, to form the very toxic metabolite CN (Mohamadin et al. 1996; Abdel-Naim and Mohamadin 2004). Therefore, it is hypothesized that HAN-induced oxidative stress and generation of ROS would result in the release of CN from HANs. Release of CN ions within the GIT epithelial cellular system may initiate and enhance oxidative stress in the cell (Mohamadin and Abdel-Naim 1999; Mohamadin et al. 2005). CN is known to induce chemical hypoxia. Such hypoxia enhances anaerobic ATP degradation and conversion of xanthine dehydrogenase to xanthine oxidase, which utilizes unmetabolized molecular oxygen to form superoxide anion. This anion is converted by superoxide dismutase to H_2O_2, which is converted to hydroxyl radicals by various electron donating metals such as iron. Alternatively, H_2O_2 is converted to hypochlorous acid by myeloperoxidase or chloroperoxidase enzymes using chloride (the detoxification product of CN^-) as a good electron donor for gastric oxidase. All of these reactions cause ROS recycling in the GIT. Therefore, the mechanisms by which HANs are bioactivated to induce GIT damage is dependent on the physiological, biochemical, and molecular functions of the gut (Fig. 3).

Because the GIT is the major target organ of HAN toxicity, the sensitivity of gastric tissue to HANs was substantiated in a dose-response study in which the effect of DBAN on GSH and GST, in male Sprague-Dawley rats, was examined (Ahmed et al. 1991a). DBAN was chosen as a model compound because it has been reported to exert the highest acute and genetic toxicities among HANs, and because it is highly reactive toward GSH conjugation *in vitro* (Ahmed et al. 1989; Deppierre et al. 1984). Although HAN-induced GSH depletion in liver and stomach is dose-dependent, the magnitude of GSH depletion in liver was less pronounced at various dose levels than in the stomach. In both organs, GST activity was decreased in a dose-dependent manner (Ahmed et al. 1991a). Because GSH activity in the GIT is neither easily induced, nor recovered once damaged, chemically-induced gut effects may predispose GIT to

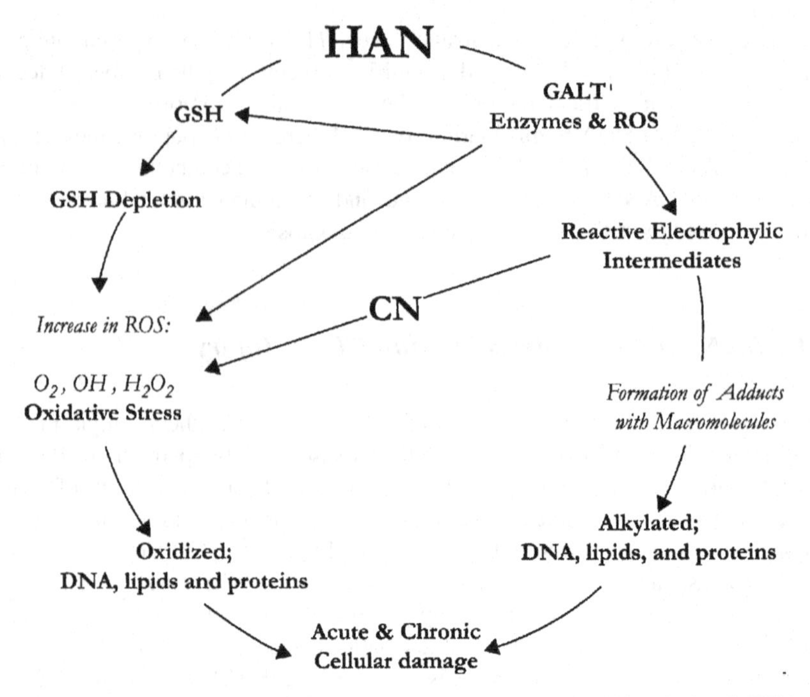

Fig. 3 Proposed biochemical pathways for halogenated acetonitrile (HAN)–induced GIT (gastrointestinal tract) injury [1]*GALT* Gut-Associated Lymphoid Tissue

necrotic and carcinogenic injury. If true, GSH depletion and GST inhibition may play roles in the GIT toxicity of HANs, particularly in the stomach.

Damage in the gut is determined by the balance between rates of formation and catabolism of ROS. The duration that ROS have tissue contact depends on the dynamic ratio of ROS/antioxidants, including GSH and other protective agents: catalase, GSH peroxidase, GSH reductase and GST. Accordingly, the mechanism of CAN's GIT toxicity was studied in the context of its effect on GSH homeostasis and impact on oxidative DNA damage in rat gastric mucosa *in vivo* (Ahmed et al. 1999; Nouraldeen et al. 1993) and *in vitro* (Mohamadin and Abdel-Naim 1999). The effect of CAN treatment on the integrity of gastric DNA was monitored by tracking the electrophoretic migration of genomic DNA on agarose gel. The impact of CAN-induced GSH depletion of gastric DNA was quantitated by measuring levels of alkyated (cyanomethyl d-guanine) and oxidized DNA bases (8-Hydroxy d-guanine; 8-OHdG) by HPLC-electrochemical detection (HPLC/EC). CAN induced a significant, dose- and time-dependent decrease (56% and 39% of control, respectively) in GSH levels in pyloric stomach mucosa at 2 and 4 hr after treatment. In addition, CAN significantly elevated gastric cyanomethyl- and 8-OH-dG levels in a dose dependent manner. Maximum levels of 8-OHdG in gastric DNA were observed at 6 hr after CAN treatment (146% of control levels). When a high dose of CAN (76 mg/kg) was used, a peak level of 8-OHdG (177% of control levels) was observed earlier (2 hr) following treatment (Ahmed et al. 1999). When CAN was incubated with gastric mucosal cells *in vitro*, a concentration-dependent CN liberation and significant decrease in cellular ATP levels were detected. The authors suggested that a mechanism for CAN-induced toxicity may be partially mediated by depletion of GSH, CN release, interruption of energy metabolism and induction of oxidative stress-damage of gastric DNA (Ahmed et al. 1999). CAN also caused a decrease in cellular GSH content and induced lipid peroxidation in gastric epithelial cells (GEC). These toxic responses were dependent on both concentration and duration of exposure. Treatment of GEC with different thiol group donors, antioxidants and iron chelators, prior to CAN exposure, protected against CAN-induced cytolethality, as indicated by measuring (lactate dehydrogenase; LDH) leakage and lipid peroxidation; this result reflects the critical role of antioxidants in CAN-induced cellular injury (Mohamadin and Abdel-Naim 1999).

5 Genotoxicity and Carcinogenesis of HANs

5.1 *Genotoxic Effects*

HANs are known to be direct-acting mutagens (WHO 1993). Although the genotoxic effects of HANs were extensively investigated, there is wide discrepancy in reported results (Bull and Robinson 1985; Meier et al. 1985b; Gee et al. 1985; Osgood and Sterling 1991; Zimmermann and Mohr 1992).

5.1.1 Prokaryotic Cells *In Vitro*

The ability of HANs to induce gene reversion in *Salmonella typhimurium* tester strains has been investigated in one of several surveys sponsored by the National Toxicology Program (Mortelmans et al. 1986). Variable results were obtained when the compounds were tested at levels up to 3,333 μg/plate +/−S9 (where S9 = 9000× g supernatant) in tester strains TA100, TA98, TA1535 and 1537. Negative findings were reported for CAN, BAN and TCAN, irrespective of the presence of molecular activation. DCAN induced gene reversion in a dose dependent manner, most markedly in strains TA100 and TA1535, irrespective of the presence of S9. The effects with DBAN were equivocal. Positive findings in one test system (in TA1535, +S9) were not matched by those in other tests.

Gene reversion and mutagenicity of HANs were investigated in a series of experiments by Bull et al. (1985). Positive results were obtained for DCAN in TA1535 and TA100, in the presence or absence of S9. All compounds were negative for gene reversion in TA1537 and TA1538; CAN, DBAN and TCAN elicited no dose-response effects in TA1535, TA98 and TA100. The results for BCAN showed similar trends to those of DCAN, though the compound's toxicity appeared to be greater than that of DCAN.

In addition, Le Curieux et al. (1995) examined the mutagenicity of HANs using the "Ames fluctuation" test. In support of the results of Bull et al. (1985), DCAN was found to induce gene reversion in the presence and absence of S9, at concentrations up to 100 μg/mL. However, positive results were also obtained for a number of other HANs, including CAN (+S9), TCAN (–S9), MBAN (+S9) and BCAN (–S9). The reversion factor (R), the maximum increase in the number of revertants/well, was calculated as -ln[(96–n)/96], where n is the number of positive wells and 96 is the number of wells per test. The following decreasing rank order for mutagenicity was demonstrated: DCAN > BCAN > BAN > TCAN > CAN, according to computed values for R. However, the authors failed to explain the apparent inconsistencies in the differing results with S9, among HANs. These findings tend to confound attempts to identify a single common mechanism for the mutagenicity of HANs, as a group.

Le Curieux et al. (1995) assessed the ability of HANs to induce DNA repair using the SOS chromotest, an experimental approach that indirectly monitors DNA damage and repair by measuring changes in the activities of β-galactosidase and alkaline phosphatase after *Escherchia coli* PQ37 isolates are incubated with test compounds. The activity of β-galactosidase serves as an indicator of DNA damage, because its gene has been fused to a DNA repair gene that is, itself, induced by a test chemicalin response to cellular DNA damage. Le Curieux et al. (1995) suggested that a compound is considered to be an SOS repair system inducer if (1) the induction factor is >1.5, (2) the β-galactosidase activity is significantly increased compared to the solvent control, (3) the induction factor/concentration graph displays a dose-response relationship, and (4) the results are reproducible. Positive results were obtained for DCAN (+S9), DBAN (–S9) and BCAN (–S9), although the responses were considered to be weak, because the maximum observed induction factor was 2.0 and ranges of effective genotoxic concentrations were narrow.

5.1.2 Eukaryotic Cells *In Vitro*

Bull et al. (1985) examined the ability of HANs to induce sister chromatid exchanges (SCEs) in Chinese hamster ovary (CHO) cells *in vitro*. Results indicated that all HANs induced SCE (+/–S9) in a dose-dependent manner. Bull et al. (1985) described the descending rank order for SCE potency as follows: DBAN > BCAN > TCAN > DCAN > CAN. These results suggest that substitution of bromine for chlorine enhances activity, and degree of halogenation promotes SCE formation.

Similar results were obtained in a recent study by Muellner et al. (2007). In this study, seven HANs: monoiodoacetonitrile (MIAN), BAN, DBAN, BCAN, CAN, DCAN and TCAN were evaluated in microplate-based CHO cell assays for chronic cytotoxicity and acute genotoxicity. Cytotoxicity was observed over the range of concentrations from 2.8 μM (DBAN) to 0.16 mM (TCAN), with a descending rank order of DBAN > MIAN > BAN > BCAN > DCAN > CAN > T CAN. In addition, HANs induced acute genomic DNA damage; the single cell gel electrophoresis genotoxicity potency ranged from 37 μM (MIAN) to 2.7 mM (DCAN) and the rank order of declining genotoxicity was MIAN > BAN > DBA N > BCAN > CAN > TCAN > DCAN.

Zimmermann and Mohr (1992) examined the ability of several chemicals, including DBAN, to induce chromosome loss in *Saccharomyces cerevisiae* D61.M, a diploid yeast strain that is heterozygous for three recessive alleles flanking the centromere of chromosome VII. The marker *cyh2* confers resistance to cycloheximide, *leu1* is a mutation causing a requirement for leucine, and *ade6* prevents the accumulation of a red pigment in cells homozygous for *ade2*. Loss of the homologous chromosome carrying the dominant markers leads to the simultaneous expression of all three recessive alleles, causing the formation of white colonies that require leucine, but which are resistant to the toxic effects of cycloheximide. DBAN was one of many compounds that induced chromosome loss in the presence of propionitrile, a well-documented inducer of chromosomal malsegregation. However, DBAN was ineffective in the absence of this inducer.

The ability of HANs to interact with, and perturb the genome of eukaryotic cells *in vitro* was also examined by Daniel et al. (1986). CCRF-CEM human lymphoblastic cells were incubated with HANs and the isolated DNA was examined for strand breaks using a DNA alkaline unwinding assay. All HANs appeared to induce strand breaks in the DNA of CCRF-CEM cells, most markedly in the case of TCAN.

5.1.3 Acellular Systems

Daniel et al. (1986) examined the capacity of ^{14}C-DCAN to bind to and/or react with double stranded calf thymus DNA or polyadenylic acid, *in vitro,* in the absence of S9. Repeated precipitation in ethanol was used to separate bound and non-bound DCAN, after which nucleic acid preparations were hydrolyzed and the fragments separated by HPLC. DCAN formed adducts with both calf thymus DNA and polyadenylic acid. Although a chromatographic peak potentially representing an adduct formed between DCAN and adenosine residues was observed, the chemical nature of the adduct was not described.

A similar experimental approach to that of Daniel et al. (1986) was used to explore the ability of BAN, CAN, DCAN and TCAN to interact with double- or single-stranded calf thymus DNA in the absence of S9 (Nouraldeen and Ahmed 1996). Fluorescence spectrophotometry was used to analyze the reaction products and provided data on the dose-response, time course, pH and temperature dependency of DNA alkylation by these HANs. In subsequent experiments, HPLC was also employed to separate adducts formed when single stranded DNA reacted with BAN. A major peak of the BAN-DNA interaction products co-eluted with synthesized

7-(cyanomethyl)guanine, suggesting that 7-(cyanomethyl)guanine is the principal product formed when DNA and BAN interact, at least *in vitro*.

5.1.4 *In Vivo* Studies

Several *in vivo* experimental studies, in which the genotoxicity of HANs has been examined, have been reported. The clastogenicity of HANs was evaluated by assessing the formation of micronuclei in polychromatic erythrocytes in mice (Bull et al. 1985) and in newt larvae (Le Curieux et al. 1995). Neither CAN, DCAN, TCAN, BCAN nor DBAN produced increases in micronuclei in mice. CAN, DCAN, TCAN, BAN, DBAN and BCAN each induced micronuclei in newt larvae. Under the limited micronuclei test protocol, no comparison of potency could be made, but the induction of micronuclei disappeared when the concentrations tested (TCAN, BAN, BCAN) were below 0.1 μg HANs/mL medium, indicating a potential threshold effect. Species differences for this effect, and the generally higher metabolic activity of mice compared to amphibians indicate that metabolism affects the degree to which HANs produce their clastogenic effect. DCAN appeared to be the most strongly mutagenic HAN, with other analogs showing fewer observations of positive effects and generally less intense effects.

Meier et al. (1985a) conducted an experiment to determine the *in vivo* genotoxic potential of HANs in Swiss CD-1 mice, as judged by the ability of the compounds to induce morphological abnormalities in the sperm heads of treated males. The mouse sperm-head assay tests the ability of chemicals to disrupt sperm morphology as a measure of mutagenic potential to a germ cell line. Animals received five daily gavage doses (0, 12.5, 25 and 50 mg/kg) of HANs in aqueous medium at 24-hr intervals, prior to sacrifice at 1, 3 or 5 week after dosing. None of HANs under investigation induced any effects on the incidence of sperm head abnormalities, when compared to controls (Meier et al. 1985a).

Lin et al. (1992) examined the ability of TCAN to bind to DNA *in vivo*, when male F344 rats were administered 1- or 2-^{14}C-TCAN in tricaprylin (0.05–0.48 mmol/kg, 1.26–7.37 mCi/mmol, 2 mL/kg) by gavage. In this model, TCAN was found to bind to both DNA and intracellular proteins in a dose-dependent manner. TCAN exhibited the highest affinity for DNA in the stomach epithelium, followed by the liver and kidney, with higher levels of bound radioactivity obtained when the radiolabel was at the C-2-position of TCAN. A chemical analysis of the bound radiolabel was not undertaken.

5.2 *Carcinogenic Effects*

Despite the fact that HANs demonstrate genotoxic activity in *in vitro* and *in vivo* bioassays, few carcinogenicity studies have been undertaken in intact animals.

5.2.1 Skin Tumor Production

The ability of HANs to act as tumor initiators in the skin of Sencar mice was examined (Bull and Robinson 1985; Bull et al. 1985). DBAN, BCAN and CAN significantly increased the yield of squamous cell papillomas and carcinomas in the skin of treated mice. HANs were applied at doses of 200, 400 or 800 mg/kg in 0.2 ml acetone to the shaved backs of mice. Six doses were applied over a period of 2 week for cumulative total doses of 1,200, 2,400 or 4,800 mg/kg. Two week after the last HAN dose, a promotion schedule was started using topical applications of 1.0 μg 12-*O*- tetradecanoylphorbol-13-acetate (TPA) three times weekly for a 20-week period. Appearance and regression of papillomas were charted weekly. Total tumor count and time to first tumor occurrence were analyzed for dose-response, with a trend test based on Cox's regression model. The study revealed that the incidence of papillomas at 24 week increased in a dose-related manner for both CAN and BCAN, and reached a maximum of 50% in the 4,800 mg/kg dose group. Also, the incidence of tumors at 24 week, in animals which received a cumulative-dose of 2,400 mg/kg of DBAN, was 65%; the incidence decreased, however, to 30% at the high-dose of 4,800 mg/kg. Apparently, DBAN caused severe ulcerations at the high-dose, which may be responsible for the decreased incidence of tumors. DCAN and TCAN induced a non-significant increase in skin tumor incidence under the same conditions. Some animals were maintained for 1 yr to determine the incidence of squamous cell carcinomas. Similar results were obtained for the incidence of squamous cell carcinoma after 1 yr, as was observed for papilloma incidence after 24 week.

Bull and Robinson (1985) examined skin carcinogenicity of HANs. Doses of 400 mg/kg for DBAN and 800 mg/kg for CAN, DCAN, TCAN and BCAN were applied three times weekly for 24 week. No increases in the incidence of skin tumors were observed in any animals receiving HANs without TPA promotion. In addition, oral dose experiments were conducted to examine the ability of HANs to act as tumor initiators in the skin of Sencar mice. A maximum dose of 50 mg/kg/d was administered six times over a 2-week period, followed by the 20-week TPA promotion schedule (Bull and Robinson 1985). There were neither significant differences in the number of tumors, or time to first tumor with any HANs in the oral dose experiment, nor any indication of other pathological effects for these compounds in topically treated animals.

5.2.2 Lung Tumor Production

Bull and Robinson (1985) observed significant increases in the incidence of lung adenomas following oral administration of a single dose of 10 mg/kg three times weekly for 8 week to female A/J mice. Dosing was initiated at 10 week of age and animals were sacrificed at 9 mon of age. Significantly elevated yields of lung adenomas were observed with CAN, TCAN, BCAN and ACN. Marginal increases in the number of lung adenomas, that did not achieve statistical significance, were observed with DCAN and DBAN. However, the authors warned that these results should be interpreted with caution because of a large variation in background rates of tumor formation in the lungs of this strain of mouse.

5.2.3 Liver Tumor Production

Based on the finding of tumor initiating activity in the mouse skin assay, Lin et al. (1986) screened CAN, BCAN, DCAN and DBAN for tumor-initiating activity in a rat liver foci bioassay. This assay involves the assessment of the induction of γ-glutamyl transpeptidase (γ-GT) positive foci in liver, an event that is a putative preneoplastic lesion. CAN was administered orally in tricaprylin at a dose of 1 mmole/kg (75.5 mg/kg) and DBAN, DCAN and TCAN at a dose of 2 mmole/kg (398, 220 and 289 mg/kg, respectively) to F344 rats, 24 hr after a 2/3 partial hepatectomy. Diethylnitrosamine (DEN) was administered as a positive control and tricaprylin was administered as a vehicle control. One week after the operation, rats started receiving 500 mg/L sodium phenobarbital (a known tumor promoter) in drinking water for 8 week. Rats were sacrificed 1 week after treatment ended. No HAN compound tested induced γ-GT positive altered foci; DEN produced a statistically significant 30-fold elevation in altered foci over vehicle control values. Although these results are negative, their predictive value is limited by the fact that only one dose level was tested, and the number of animals was limited (7–10 per group).

6 Mechanism of Action

6.1 General Mechanistic Pathways

The precise mechanism by which HANs produce acute or chronic toxicity is currently not defined. However, as mentioned before, HANs are metabolized to reactive intermediates which can interact with structural and functional macromolecules (proteins and lipids) to produce toxicity. The proposed metabolic pathway, shown in Fig. 1, illustrates the chemical mechanisms of HANs biotransformation to highly reactive intermediates that play a role in HAN-induced toxicity and carcinogenicity. In the upper pathway (Fig. 1) the electron withdrawing (and good leaving) halogroup(s) may undergo nucleophilic displacement, possibly at sulfhydryl and other nucleophilic sites (Rnus) in cellular molecules, to form cyanomethyl derivatives. This reaction would lead to various toxic interactions, the nature of which would depend on the type of biological nucleophiles involved. As an example, interaction with informational macromolecules such as nucleic acids or the enzymes regulating their functions, may lead to genetic or epigenetic aberrations such as mutagenesis and carcinogenesis.

The lower pathway (Fig. 1) involves oxidative biotransformation and CN^- release, which requires metabolic activation of the molecule by oxygen insertion between C-H bond, before cleavage of the cyano group. The ultimate result is the formation of highly reactive haloaldehyde or ketone compounds. If produced in large quantities, these compounds may overwhelm the detoxification mechanisms of the cell, potentially resulting in toxic molecular interactions constituting key components of the toxic mechanism for HANs.

This pathway (Fig. 1) in mammals is similar to the one that results in microbial transformation of ACN and CAN (Castro et al. 1996). The first step in oxidative and hydrolytic transformations of ACN and CAN is insertion of an oxygen into the C-H bond to yield the corresponding cyanohydrins. With ACN, the loss of HCN from the initially formed cyanohydrin (HOCHCN) produces formaldehyde. The latter is rapidly converted to formate and CO_2. With CAN, the first formed cyanohydrin (HOCHClCN) may lose HCN to produce formyl chloride, or HCl may be lost to yield formyl CN. The generation of CO, in addition to formate, implicates formyl chloride and/or formyl CN as intermediates (Castro et al. 1996). The formation of multiple reactive metabolites complicates understanding the mechanism(s) of HAN-induced toxicity. Further, the presence of multiple metabolites, and dearth of quantitative data describing the formation of these metabolites, are barriers to employing biomarkers of HAN exposure in a quantitative manner.

The results of several *in vivo* and *in vitro* studies on the interaction of HANs with GSH agree with the hypothesis that HANs possess two reactive centers with which nucleophilic agents can react. The reaction at the first center involves initial displacement of a halogen atom at the α-carbon atom prior to binding; the second reaction involves binding via addition to the cyano-group triple bond. It was found that HANs react with 4-(*p*-nitrobenzyl)-pyridine by displacing a halogen atom. The relative rates for such displacement are: DBAN > BCAN > MCAN > DCAN > TC AN (Daniel et al. 1986). In the reaction of HANs with GSH, a similar order of reactivity was observed, except with TCAN. The electron-withdrawing substituent on the α-carbon atom increases the reaction rate of nucleophilic addition on the cyano group carbon. Thus, TCAN is the most reactive HAN toward nucleophilic addition to the bond in the cyano group, in contrast to the nucleophilic replacement of halogen on the α-carbon atom. Therefore, GSH may react with DBAN, BCAN and MCAN by displacing a halogen atom on the α-carbon, and react with TCAN by addition to the CN carbon atom.

Accordingly, the toxic and genotoxic consequence of HANs interaction with cellular macromolecules may be significantly affected by the presence of GSH. For mono- and tri-haloacetonitriles, GSH conjugation should result in detoxification, because elimination of reactive electrophiles is enhanced. There is evidence that the toxicity and genotoxicity of HANs may not be fully manifested until the cellular GSH pool is depleted (Abdel-Aziz et al. 1993). For dihaloacetonitriles, GSH conjugation only results in detoxification when both halogens are displaced. If only one is displaced, GSH conjugation can become an activation pathway, because the resulting intermediate (halothioether) is a highly reactive electrophile. There are many examples of dihaloalkanes being activated by GSH conjugation (Woo et al.

1988). The relative importance of the GSH activation pathway for dihaloacetonitriles remains to be evaluated.

6.2 Oxidative Stress as a Signaling Pathway for HANs-Induced Adverse Effects

Evidence supports the role of oxidative stress in HANs-induced toxicity. *In vitro* and *in vivo* studies indicate that HANs interact with cellular nucleophiles, which have an important role in the antioxidant defense system of the cells, particularly GSH and GST (Ahmed and Hussein 1987; Ahmed et al. 1991a; Deppierre et al. 1984; Lin et al. 1992). GSH and similar enzymes protect the cell against electophilic, necrotic and carcinogenic agents. Mechanisms that deplete GSH in tissues result from (a) depletion from formation of GSH conjugates catalyzed by GST (Chassaud 1979); (b) an increased rate of oxidation of GSH to GSSG catalyzed by the enzyme GSH peroxidase (Richman et al. 1973); or (c) a decreased rate of biosynthesis of GSH (Fu et al. 1990).

To explore the mechanism of interaction of HANs with the GSH system, ACN and seven of its halogenated derivatives were evaluated for their interaction with hepatic cytosolic GST activity *in vitro,* using CDNB as the substrate (Lin and Guian 1989). Increasing concentrations of ACN, monofluoroacetonitrile (MFAN), CAN, and BAN up to 10 mM failed to produce 50% inhibition of the activity of GST enzyme. However, DCAN, TCAN, DBAN and MIAN were inhibitors with inhibition rate constant (I_{50}) values of 2.49, 0.34, 0.82 and 4.44 mM, respectively. The inhibitory effect of HANs on hepatic GST activity toward CDNB was mixed. The inhibitory effect on hepatic GST activity for DCAN, DBAN and TCAN were found to be reversible, and the activity was completely recovered after dialysis of the inhibited enzyme. In contrast, MIAN inhibited GST activity in an irreversible manner. HANs-induced inhibition of hepatic GST activity *in vitro* is consistent with that observed *in vivo* (Ahmed et al. 1991a). Because HANs induced reversible inhibition of hepatic GST activities, this effect may lead to decreased detoxification of HANs and other electrophilic chemicals (Ahmed et al. 1989).

The rank-order of HANs reactivity was also quantified by assessing the ability of CAN, DCAN, TCAN, BCAN and DBAN to react with GSH *in vitro* (Lin and Guian 1989). Results indicated that DBAN and BCAN were the most potent electrophiles tested, whereas DCAN appeared not to react with GSH. The addition of microsomal enzymes (including cytochrome P-450 forms) resulted in an increased depletion of GSH in incubations containing CAN, but GSH reactivity was diminished in incubations containing either TCAN or DBAN. This may indicate that CAN is metabolized by microsomal enzymes to a more reactive intermediate than are TCAN or DBAN.

As mentioned, HANs decreases the GST-catalyzed conjugation of GSH with CDNB by three mechanisms: (a) by depleting GSH through a direct reaction possibly important for DBAN and possibly (because of their high reactivity toward

GSH); (b) by competing with CDNB as a substrate for conjugation with GSH (among HANs, only CAN conjugates with GSH); and (c) by binding to GST, thus inactivating the enzyme. HANs may inactivate GST by interaction with thiol groups, amino groups or other nucleophilic centers in the protein through replacement of a halogen atom (CAN, DBAN, BCAN) or reaction at a cyano group (TCAN and DCAN) (Castro et al. 1996; Chance et al. 1979).

To further evaluate the mechanisms of HAN-induced oxidative stress, several studies were conducted to investigate the impact of HAN-induced GSH depletion and CN release on cytotoxicity and programmed cell death (apoptosis). Ahmed et al. (2000) evaluated the effect of various concentrations of DCAN (100–400 μM) on the activity of immortalized monocyte macrophage cells (RAW 264.7). DCAN-induced oxidative stress was characterized by the production of reactive oxygen intermediates (ROI). The ratios of intracellular GSH/GSSG was assessed and used as a biomarker for oxidative stress. The secretion of TNF-α (tumor necrosis factor alpha) was assessed, because macrophages are known to secrete it in response to cellular oxidative stress. Electrophoretic detection of DNA degradation and light microscopy were utilized to characterize DCAN-induced apoptosis. LDH leakage and trypan blue exclusion were used as markers of cellular necrosis. Following exposure of macrophages to DCAN (200 μM and 400 μM), a significant increase in levels of intracellular GSSG and ROI were detected. Increased secretion of TNF-α was also observed. Elecrophoresis of treated-cell genomic DNA indicated a dose-dependent increase in degradation of genomic DNA. Results of morphological studies showed that exposure of RAW cells to 100 μM or 200 μM DCAN induced apoptotic cell death. At higher concentrations (400 μM), however, a significant increase in LDH leakage and decrease in cell viability (55% of control) indicative of cellular necrosis was observed. The authors concluded that DCAN induced dose dependent apoptosis or necrosis in RAW cells; this cellular damage may result from disturbance of intracellular redox status and initiation of ROI-mediated oxidative mechanisms (Ahmed et al. 2000). The mechanism of DBAN-induced apoptosis and cytotoxicity in human lymphoid leukemia and rat intestinal epithelial cells (Jacob et al. 2006) were investigated. Results indicate that there is a concentration and time dependent increase in DBAN-induced apoptosis in human lymphoid leukemia cells, as indicated by Annexin V binding and TUNEL (terminal deoxynucleotidyl transferase biotin-dUTP Nick End Labeling) assays. Microscopic evaluation of Hoechst- or Giemsa-stained cells showed increased nuclear fragmentation and apoptic features. At a higher dose (10 μM) and at longer time periods (24–48 hr) DBAN induced severe cellular membrane damage. Moreover, morphological evaluation of Giemsa-stained rat intestinal epithelial cells revealed various features of apoptosis including cell shrinkage, and nuclear and chromatin condensation at 5–10 μM concentrations of DBAN. Severe nuclear fragmentation was observed at higher DBAN concentrations (20–50 μM). Apoptosis was further verified by labeling of cells with TUNEL which showed increased uptake of the dye by apoptotic nuclei (AE Ahmed personal communication).

To examine the mechanism of DBAN- induced DNA damage, mouse embryonic fibroblasts were exposed to various concentrations (5–40 μM) of DBAN for

10–240 min (AE Ahmed, personal communication). Phase contrast microscopic and morphologic evaluation of DBAN-treated cells indicated features of apoptosis at low exposure levels, and severe membrane damage at high concentrations of DBAN. The data show a concentration- and time-dependent (up to 40 min) depletion (up to 80%) of total GSH. At later times, however, GSH concentrations rebounded. Lipid peroxidation, measured as malondialdehyde, was significantly increased (up to fivefold). The ability of fibroblasts to undergo DNA repair was studied using base excision repair assay (BER), a predominant DNA repair mechanism involving single nucleotide gap-filling DNA synthesis. Data showed that BER activity was up regulated in cells exposed to 0.1–5 μM DBAN for 10 min, and was inhibited at higher concentrations and longer exposure periods. This phenomenon is known as hormesis.

In another acelluar system, HAN-induced strand breaks in plasmid DNA, was evaluated *in vitro*. A unique *in vitro* assay was developed to investigate the scope of oxidative mechanisms of HANs-induced DNA damage (AE Ahmed, unpublished findings). DNA strand breaks were assayed by measuring the ability of HANs to convert super-coiled double stranded p-Bluescript DNA plasmid (SC) to open relaxed circular (C) or linear (L) forms. Results showed that, when DCAN was added to the incubation mixture, there was a significant time- and concentration- increase in the conversion of super coiled (32–55% of the total) DNA plasmid to circular form. A concentration-dependent increase in the levels of 8-OHdG, the marker for oxidative DNA damage, was observed following HPLC/EC analysis of DCAN-treated plasmid DNA hydrolysates. Antioxidants such as GSH, melatonin, catalase and superoxide dismutase and vitamins E and C protected against DCAN-induced DNA damage. This study supports the hypothesis of HAN-induced oxidative DNA damage as a possible mechanism for DCAN toxicity (AE Ahmed, unpublished findings).

7 Comparison of the Relative Chemical and Biological Activities of HANs

Several comparative HAN studies are summarized in Table 3. To facilitate comparison of results, data are normalized to the response expressed by CAN. Using Pearson Multiple Correlation statistics, the relative activities of five HANs for metabolism, alkylation potential, DNA strand breaks and carcinogenic potential were compared. The endpoints included inhibition of DMN-DM, the extent of excretion of SCN (Pereira et al. 1984), the ability to alkylate 4-(*p*-nitrobenzyl) pyridine (NBP) as an index for alkylation potential, DNA strand breakage potential in CCRF-CEM cells as an index for their interaction with DNA (Daniel et al. 1986) and skin tumor induction (Bull and Robinson 1985).

As shown in the Table 3, CAN exhibited a relatively high alkylation potential and rate of metabolism as measured by excretion of SCN, compared to its ability to inhibit *in vitro* microsomal DMN-DM activity, and to induce DNA strand breaks. In contrast,

Table 3 Comparison of the relative chemical and biological activities of HANs[a]

	Metabolism		Genotoxicity		Carcinogenesis
HANs	DMN-DM inhibition[b]*	Extent metabolized[c]	Alkylation potential[d]	DNA strand breaks[e]	Skin tumor induction[f]
CAN	1	1	1	1	1
DCAN	450	0.7	0.07	2.1	0.72
TCAN	450	0.2	0.01	37	0.88
DBAN	3,000	0.5	6.2	3.4	0.5
BCAN	2,300	0.9	2.2	6.3	0.97

[a] To facilitate comparison of results, data for all halogenated acetonitriles (HANs) shown are normalized to CAN (CAN = 1); Table data adapted from Daniel et al. 1986; Lin et al. 1986, and Muellner et al. (2007)

[b] Expressed as the relative *in vitro* inhibition of rat hepatic microsomal dimethylnitrosamine-demethylase (DMN-DM) (Lin et al. 1986)

[c] Extent of metabolism was measured as the percentage of the dose excreted as urinary thiocyanate (SCN) and expressed relative to CAN (Pereira et al. 1984)

[d] Alkylation potential is expressed as the relative ability to alkylate 4-(*p*-nitrobenzyl)pyridine (NBP) (Daniel et al. 1986)

[e] DNA strand breakage potential in CCRF-CEM cells was determined by the alkaline unwinding method (Daniel et al. 1986)

[f] Expressed as the total number of skin tumors per mouse after a topical dose of 2.4 g/kg of the HAN, followed by treatment with 12-*O*-tetradacanoylphorbol-13-acetate for 20 week; data from Bull et al. (1985); ranking from Daniel et al. (1986)

* The *in vitro* inhibition of rodent microsomal DMN-DM is significantly correlated with alkylation potential ($r = 0.88$, $P < 0.05$)

the alkylation potential and the metabolism to SCN of TCAN were low compared to its propensity to inhibit DMN-DM activity and cause DNA strand breaks.

The discrepancy between the akylation potential and the extent of metabolism to SCN of CAN and TCAN, and their ability to inhibit DMN-DM and cause DNA damage, may result from the much greater chemical reactivity of the proposed TCAN metabolites, phosgene and cyanoformyl chloride compared to the proposed CAN metabolite, formaldehyde. Therefore, although TCAN degrades less to CN than CAN, phosgene and cyanoformyl chloride formation is expected to be more potent as inhibitors of DMN-DM, and as inducers of DNA strand-breaks than is formaldehyde. It is also possible that, compared to TCAN, the higher reactivity (alkylation potential) of CAN may indicate that it is too reactive to reach target intracellular macromolecules (Pereira et al. 1984). Thus, the relative biologic activity of HANs would result from the activity of both the parent chemicals and their reactive metabolites.

Although there is a significant correlation between DNA alkylation potential of HANs and their inhibitory effect on microsomal DMN-DM, a non significant correlation was observed between their alkylation potential and their ability to produce DNA strand breaks in cultured cells. The contrasting results between the binding of HANs to NBP, and their ability to cause DNA strand breaks or inhibit DMN-DM activity may be explained on the basis of molecular interactions. The NBP reaction probably occurs via nucleophilic attack by NBP on the HAN with resultant halogen displacement and binding; whereas the DMN-DM inhibition and DNA strand breaks occur both by displacement of halogen and by attack on the cyano group (Lin et al. 1986).

As mentioned before, HANs are absorbed systemically and are converted to toxic metabolites as demonstrated by their excretion as SCN and their inhibition of DMN-DM activity in rat liver. However, HANs failed to produce any detectable amounts of DNA adducts in rat liver following oral administration (Lin et al. 1986), and failed to initiate γ-GT foci in the rat liver foci bioassay. The inability to demonstrate systemic activity for HANs may result from toxic metabolic compounds failing to reach the target site and/or from their rapid *in vivo* detoxification. The tumor-initiating activity of HANs topically applied to mouse skin indicates that a carcinogenic hazard may exist at the site of application. Thus, the carcinogenic risk posed by HANs in drinking water may be limited to the GIT, and other organs in direct contact with them, such as skin (during showering) and lung (by inhalation). Further investigations are required to determine if HANs are gastrointestinal carcinogens.

8 Summary

The haloacetonitriles (HANs) exist in drinking water exclusively as byproducts of disinfection. HANs are found in drinking water more often, and in higher concentrations, when surface water is treated by chloramination. Human exposure occurs through consumption of finished drinking water; oral and dermal contact also occurs, and results from showering, swimming and other activities.

HANs are reactive and are toxic to gastrointestinal tissues following oral administration. Such toxicity is characterized by GSH depletion, increased lipid peroxidation, and covalent binding of HAN-associated radioactivity to gut tissues. The presence of GSH in cells is an important protective mechanism against HAN toxicity; depletion of cellular GSH results in increased toxicity. Some studies have demonstrated an apparently synergistic effect between ROS and HAN administration, that may help explain effects observed in GI tissues. ROS are produced in gut tissues, and *in vitro* evidence indicates that ROS may contribute to the degradation and formation of reactive intermediates from HANs. The rationale for ROS involvement may involve HAN-induced depletion of GSH and the role of GSH in scavenging ROS. In addition to effects on GI tissues, studies show that HAN-derived radiolabel is found covalently bound to proteins and DNA in several organs and tissues. The addition of antioxidants to biologic systems protects against HAN-induced DNA damage. The protection offered by antioxidants supports the role of oxidative stress and the potential for a threshold in HAN-induced toxicity. However, additional data are needed to substantiate evidence for such a threshold.

HANs are readily absorbed from the GI tract and are extensively metabolized. Elimination occurs primarily in urine, as unconjugated one-carbon metabolites. Evidence supports the involvement of mixed function oxidases, the cytochrome

P450 enzyme family and GST, in HAN metabolism. Metabolism represents either a detoxification or bioactivation process, depending on the particular HAN and the enzyme involved. HANs can inhibit CYP2E1-mediated metabolism, an effect which may be dependent on a covalent interaction with the enzyme. In addition, HAN compounds inhibit GST-mediated conjugation, but this effect is reversible upon dialysis, indicating that the interaction does not represent covalent binding.

No subchronic studies of HAN toxicity are available in the literature. However, studies show that HANs produce developmental toxicity in experimental animals. The nature of developmental toxicity is affected by the type of administration vehicle, which renders interpretation of results more difficult. Skin tumors have been found following dermal application of HANs, but oral studies for carcinogenicity are negative. Pulmonary adenomas were increased following oral administration of HANs, but the A/J strain of mice employed has a characteristically high background rate of such tumors. HANs interact with DNA to produce unscheduled DNA repair, SCE and reverse mutations in *Salmonella*. HANs did not induce micronuclei or cause alterations in sperm head morphology in mice, but did induce micronuclei in newts. Thus, there is concern for the potential carcinogenicity of HANs. It would be valuable to delineate any relationship between the apparent threshold for micronuclei formation in newts and the potential mechanism of toxicity involving HAN-induced oxidative stress. Dose-response studies in rodents may provide useful information on toxicity mechanisms and dose selection for longer term toxicity studies.

Additional studies are warranted before drawing firm conclusions on the hazards of HAN exposure. Moreover, additional studies on HAN-DNA and HAN-protein interaction mechanisms, are needed. Such studies can better characterize the role of metabolism in toxicity of individual HANs, and delineate the role of oxidative stress, both of which enhance the capacity to predict risk. Most needed, now, are new subchronic (and chronic) toxicity studies; the results of such well-planned, controlled, conducted, interpreted and published investigations would be valuable in establishing margins of safety for HANs in human health risk assessment.

Acknowledgements Work conducted in the laboratory of Dr Ahmed E. Ahmed was supported by U.S. EPA and the National Institute for Environmental Health Sciences / National Institutes of Health.

References

Abdel-Aziz AH, Abdel-Rahman SZ, Nouraldeen AM, Shouman SA, Loh JP, Ahmed AE (1993) Effect of glutathione modulation on molecular interaction of [^{14}C]-Chloroacetonitrile with maternal and fetal DNA in mice. Reprod Toxicol 7:263–272.

Abdel-Naim AB, Mohamadin AM (2004) Myeloperoxidase-catalyzed oxidation of chloroacetonitrile to cyanide. Toxicol Lett. 146(3):249–57.

Ahmed AE, Hussein GI (1987) Studies on the mechanism of haloacetonitrile acute toxicity: interaction of dibromoactonitrile (DBAN) with glutathione (GSH) and glutathione-S-transferase (GST) in rats. Toxicologist 7:452.

Ahmed AE, Soliman SA, Loh JP, Hussein GI (1989) Studies on the mechanism of haloacetonitriles toxicity: inhibition of rat hepatic glutathione-S-transferase *in vitro*. Toxicol Appl Pharmacol 100:271–279.

Ahmed AE, Hussein GI, Loh JP, Abdel-Rahman SZ (1991a) Studies on the mechanism of haloacetonitrile-induced gastrointestinal toxicity: interaction of dibromoactonitrile with glutathione and glutathione-S-transferase in rats. J Biochem Toxicol 6(2):115–121.

Ahmed AE, Jacob S, Loh JP (1991b) Studies on the mechanism of haloacetonitriles toxicity: whole-body autoradiographic distribution of 2-[^{14}C]-chloroacetonitrile in rats. Toxicology 67(3):279–302.

Ahmed AE, Loh JP, Ghanayem B, Hussein G (1992) Studies on the mechanism of acetonitrile toxicity. I: whole body autoradiographic distribution and macromolecular interaction of 2-^{14}C-acetonitrile in mice. Pharmacol Toxicol 70(5):322–330.

Ahmed AE, Jacob S, Nouraldeen AM (1999) Chloroacetonitrile (CAN) induces glutathione depletion and 8-hydroxylation of guanine bases in rat gastric mucosa. J Biochem Mol Toxicol 13(3–4):119–126.

Ahmed AE, Aronson J, Jacob S (2000) Induction of oxidative stress and TNF-alpha secretion by dichloroacetonitrile, a water disinfectant byproduct, as possible mediators of apoptosis or necrosis in murine macrophage cell line (RAW). Toxicol In Vitro 14(3):199–210.

Ahmed AE, Campbell G A, Jacob S (2005a) Neurological impairment in fetal mouse brain by drinking water disinfectant byproducts. Neurotoxicology 26:633–640.

Ahmed AE, Jacob S, Campbell GA, Harirah HM, Perez-Polo JR, Johnsone KM (2005b) Fetal origin of adverse pregnancy outcome: the water disinfectant by-product chloroacetonitrile induces oxidative stress and apoptosis in mouse fetal brain. Dev Brain Res 159:1–11.

Aschengrau A, Zierler S, Cohen A (1989) Quality of community drinking water and the occurrence of spontaneous abortion. Arch Environ Health 44:283–290.

Bhat H, Ahmed AE, Ansari GA (1990) Toxicokinetics of monochloro acetic acid: a whole body autoradiograph study. Toxicology 63:35– 43.

Bieber TI, Trehy ML (1983) Dihaloacetonitriles in chlorinated natural waters. In: Jolley RL, Brungs WA, Cotruvo JA, Cumming RB, Mattice JS, Jacobs VA (eds) Water chlorination: environmental impact and health effects, Vol 4. Ann Arbor Science Publishers, Ann Arbor, MI. pp 85–96.

Bove F, Shim Y, Zeitz P (2002) Drinking water contaminants and adverse pregnancy outcomes. A Rev Environ Health Perspect 110(Suppl. 1):61–74.

Bruchet A, Mutsumi Y, Duguet JP, Mallevialle J (1985) Characterization of total halogenated compounds during water treatment processes. In: Jolley RL, Bull RJ, Davis WR, Katz S, Roberts MH, Jr, Jacobs VA (eds) Water chlorination: chemistry, environmental impact and health effects, Vol 5. Lewis Publishers, Chelsea, MI. pp 1165–1184.

Bull RJ, Robinson M (1985) Carcinogenic activity of haloacetonitriles and haloacetone derivatives in the mouse skin and lung In: Jolley RL, Bull RJ, Davis WP, Katz S, Roberts MH, Jr, Jacobs VA (eds) Water chlorination: chemistry, environmental impact and health effects, Vol 5. Lewis Publishers, Chelsea, MI. pp 221–227.

Bull RJ, Meier JR, Robinson M, Ringhand HP, Laurie RD, Stober JA (1985) Evaluation of mutagenic and carcinogenic properties of chlorinated acetonitriles: by-products of chlorination. Fund Appl Toxicol 5:1065–1074.

Castro CE, O'Shea SK, Wang W, Bartnicki EW (1996) Biodehalogenation: oxidative and hydrolytic pathways in the transformations of acetonitrile, chloroacetonitrile and chloroacetamide by *Methylosinus trichosporium* OB-3b. Environ Sci Technol 30(4):1180–1184.

Chance B, Sies H, Boveris A (1979) Hydroperoxide metabolism in mammalian organs. Physiol Rev 59(3):527–604.

Chassaud LF (1979) The role of glutathione and glutathione-S-transferases in the metabolism of chemical carcinogens and other electrophilic agents. Adv Cancer Res 29:175–274.

Chernoff N, Kavlock RJ (1982) An *in vivo* teratology screen utilizing pregnant mice. J Toxicol Environ Health. 10:541–50.

Christ SA, Read EJ, Stober JA, Smith MK (1996) Developmental effects of trichloroacetonitrile administered in corn oil to pregnant Long-Evans rats. Toxicol Ind Health 47:233–247.

Coleman WE, Munch JW, Kaylor WH, Streicher RP Ringhand HP, Meier JR (1984) Gas chromatography/mass spectrometry analysis of mutagenic extracts of aqueous humic acid: a comparison of the by-products to drinking water contaminants. Environ Sci Technol 18:674–681.

Cotton RT, Walkden HH (1968) The role of sorption in the fumigation of stored grain and cereal products. J Kansas Entomol Soc 17:98.

Daniel FB, Schenck KM, Mattox JK, Lin ELC, Haas DL, Pereira MA (1986) Genotoxic properties of haloacetonitriles: drinking water by-products of chlorine disinfection. Fund Appl Toxicol 6:447–453.

Deppierre JW, Seidegard J, Mongenstern R, Balk L, Meijer J, Astrom A, Norelius I, Ernester L (1984) Induction of cytosolic glutathione transferase and microsomal epoxide hydrolase activities in extrahepatic organs of the rat by phenobarbital, 3-methylchloranrene and trans-stilbene oxide. Xenobiotica 14(4):295–301.

Doyle TJ, Zheng W, Cerhan JR, Hong CP, Sellers TA, Kushi LH, Folsom AR (1997) The association of drinking water source chlorination byproducts with cancer incidence among post-menopausal women in Iowa: a prospective cohort study. Am J Pub Health 87(7):1168–1176.

Fu LJ, Johnson EM, Newman LM (1990) Prediction of the developmental toxicity hazard of halogenated drinking water disinfection byproducts tested by the *in vitro* hydra assay. Regul Toxicol Pharmacol 11(3):213–219.

Gee P, Sommers CH, Melick AS, Gidrol XM, Todd MD, Burris RB, Nelson ME, Klemm RC (1985) Comparison of response of base-specific *Salmonella* tester strains with the traditional strains for identifying mutagens: the results of a validation study. Mutat Res 412:115–130.

George EL, Zenick H, Manson J, Smith, MK (1985) Developmental studies of acetonitrile and haloacetonitriles in the Long-Evans rat. Toxicologist 5:458.

Graves CG, Genevieve MM, Tardiff RG (2001) Weight of evidence for an association between adverse reproductive and developmental effects and exposure to disinfection byproducts: a critical review. Toxicol Pharmacol 34:103–124.

Grisham MB, Granger DN (1988) Neutrophil-mediated mucosal injury. Role of reactive oxygen metabolites. Dig Dis Sci 33 (Suppl. 3):6S–15S.

Hayes JR, Condie LW Jr., Borzelleca JF (1986) Toxicology of haloacetonitriles. Environ Health Perspect 69:183–202.

IARC (International Agency for Research on Cancer) (1991) Halogenated acetonitriles. In: IARC monographs on the evaluation of carcinogenic risks to humans. Chlorinated drinking water; chlorination by-products; some other halogenated compounds, cobalt and cobalt compounds, Vol 52. World Health Organization, International Agency for Research on Cancer, Lyon. pp 269–296.

Jacob S, Abdel-Aziz AA, Shouman S, Ahmed AE (1998) Effect of glutathione modulation of the distribution and transplacental uptake of 2-[^{14}C]- chloroacetonitrile (CAN). Quantitative whole-body antoradiograhic study in pregnant mice. Toxicol Ind Health 14(4):533–546.

Jacob S, Kaphalia BS, Jacob N, Ahmed AE (2006) The water disinfectant byproduct dibromoacetonitrile induces apoptosis in rat intestinal epithelial cells: possible role of redox imbalance. Toxicol Mech Methods 16:227–234.

Kim H, Shim J, Lee S (2002) Formation of disinfectant byproducts in chlorinated swimming pool water. Chemosphere 46(1):123–130.

King WD, Dodds L, Allen AC (2000) Relation between stillbirth and specific chlorination by-products in public water supplies. Environ Health Perspect 108:883– 886.

Klotz JB, Pyrch LA (1999) Neural tube defects and drinking water disinfection by products. Epidemiology 10:383–390.

Kramer MD, Lynch CF, Isacson P, Hanson JW (1992) The association of waterborne chloroform with intrauterine growth retardation. Epidemiology 3:407–413.

Krasner SW, McGuire MJ, Jacangelo JG, Patania NL, Reagan KM, Aieta EM (1989) The occurrence of disinfection byproducts in United-States drinking water. J Am Water Works Assoc 81:41–53.

Krenitsky TA, Tuttle JV, Cattau EL, Wang P (1974) A comparison of the distribution and electron acceptor specialties of xanthine oxidase and aldehyde oxidase. Comp Biochem Physiol 49B:687–703.

Le Curieux F, Giller S, Gauthier L, Erb F, Marzin D (1995) Study of the genotoxic activity of six halogenated acetonitriles using the SOS chromotest, the Ames-fluctuation test and the newt micronucleus test. Mutat Res 341:289–302.

Lin ELC, Guian CW (1989) Interaction of haloacetonitriles with glutathione and glutathione-*S*-transferase. Biochemical Pharmacol 38:685–688.

Lin ELC, Daniel FB, Herren-Freund SL, Pereira MA (1986) Haloacetonitriles: metabolism, genotoxicity, and tumor-initiating activity. Environ Health Perspect 69:67–71

Lin ELC, Reddy TV, Daniel FB (1992) Macromolecular adduction by trichloroacetonitrile in Fischer 344 rat following oral gavage. Cancer Lett 62:1–9.

Magnus P, Jaakkola JJK, Skrondal A, Alexander J, Becher G, Krogh T, Dybing E (1999) Water chlorination and birth defects.Epidemiology 10:513–517.

Matt J (1968) U.S. patent number 3,608,084. Available at http://www.uspto.gov/patft/index.html.

McGuire MJ, McLain JL, Obolensky A (2002) Information Collection Rule Data Analysis; American Water Works Association Research Foundation and AWWA: Denver, CO.

McKinney JD, Maurer RR, Hass JR, Thomas RO (1976) Possible factors in the drinking water of laboratory animals causing reproductive failure. In: Keith LH (ed) Identification and analysis of organic pollutants in water. Ann Arbor Science, Ann Arbor, MI. pp 417–432.

Meier JR, Bull RJ, Stober IA, Cimino MC (1985a) Evaluation of chemicals used for drinking water disinfection for production of chromosomal damage and sperm-head abnormalities in mice. Environ Mutagen 7:201–211.

Meier JR, Ringhand HP, Coleman WE, Munch JW, Streicher RP, Kaylor WH, Schenck KM (1985b) Identification of mutagenic compounds formed during chlorination of humic acid. Mutat Res 157:111–122.

Mink DL, Coleman WE, Munch JW, Kaylor WH, Ringhand HP (1983) *In vivo* formation of halogenated reaction products following peroral sodium hypochlorite. Bull Environ Contam Toxicol 30:394–399.

Mohamadin AM (2001) Possible role of hydroxyl radicals in the oxidation of dichloroacetonitrile by Fenton-like reaction. J Inorg Biochem 84(1–2):97–105.

Mohamadin AM, Abdel-Naim AB (1999) Chloroacetonitrile-induced oxidative stress and toxicity in rat gastric epithelial cells. Pharmacol Res 40:377–383.

Mohamadin AM, El-Zahaby MH, Ahmed AE (1996) Acrylonitrile oxidation and cyanide release in a cell free system catalyzed by a Fenton-like reaction. Toxicologist 30:238.

Mohamadin AM, El-Demerdash E, El-Beshbishy HA, Abdel-Naim AB (2005) Acrylonitrile-induced toxicity and oxidative stress in isolated rat colonocyte. Environ Toxicol Pharmacol 19:371–377.

Mortelmans K, Haworth S, Lawlor T, Speck W, Tainer B, Zeiger E (1986) Salmonella mutagenicity tests: II. Results from the testing of 270 chemicals. Environ Mutagenesis 8 Suppl 7:1–119.

Moudgal CJ, Lipscomb JC, Bruce RM (2000) Potential health effects of drinking water disinfection byproducts using quantitative structure activity relationship. Toxicology 147(2):109–131.

Muellner MG, Wagner ED, McCalla K, Richardson SD, Woo YT, Plewa MJ (2007) Haloacetonitriles vs. regulated haloacetic acids: are nitrogen-containing DBPs more toxic? Environ Sci Technol 41:645–651.

Mughal FH (1992) Chlorination of drinking water and cancer: a review. J Environ Pathol Toxicol Oncol 11(5–6):287–292.

Muller-Pillet V, Joyeux M, Ambroise D, Hartemann P (2000) Genotoxic activity of five haloacetonitriles: comparative investigations in the single cell gel electrophoresis (Comet) assay and the Ames-fluctuation test. Environ Mol Mutagen 36:52–58.

Nouraldeen AM, Ahmed AE (1996) Studies on the mechanisms of haloacetonitrile-induced genotoxicity. IV: *in vitro* interaction of haloacetonitriles with DNA. Toxicol In Vitro 10:17–26.

Nouraldeen AM, Abdel-Rahman SZ, Ahmed AE (1993) Genotoxicity and cell proliferative activity of the drinking water disinfectant byproducts haloacetonitriles (HAN) in rat stomach mucosa. Toxicologist 13:811.

Oliver BG (1983) Dihaloacetonitriles in drinking water: algae and fulvic acids as precursor. Environ Sci Technol 17(2):80–83.

Osgood C, Sterling D (1991) Dichloroacetonitrile, a by-product of water chlorination, induces aneupoloidy in Drosophila. Mutat Res 261(2):85–91.

Otson R (1987) Purgeable organics in Great Lakes raw and treated water. Int J Environs Anal Chem 31:41–53.

Pereira MA, Lin L-HC, Mattox JK (1984) Haloacetonitrile excretion as thiocyanate and inhibition of dimethyl-nitrosamine demethylase: a proposed metabolic scheme. J Toxicol Environ Health 13:633–641.

Pourmoghaddas H, Stevens AA, Kinman RN, Dressman RC, Moore LA, Ireland JC (1993) Effect of bromide ion on formation of HANs during chlorination. JAWWA 85:82–87.

Raymer JH, Pellizzari E, Childs B, Briggs K, Shoemaker JA (2000) Analytical methods for water disinfectant byproducts in foods and beverages. J Exp Anal Environ Epidemiol 10(6 Pt. 2): 808–815.

Reding R, Fair PS, Sharp CJ, Brass HJ (1989) Measurement of dihaloacetonitriles and chloropicrin in US drinking waters. In: Disinfection by-products: current perspectives. American Water Works Association, Denver, CO. pp 11–22.

Richman PG, Orlowski M, Miester A (1973) Inhibition of gamma-glutamyl cysteine synthetase by L-methionene-*S*-sulfoximine. J Biol Chem 248(19):6684–6690.

Roby MR, Carle S, Pereira MA, Carter DE (1986) Excretion and tissue disposition of dichloroacetonitrile in rats and mice. Environ Health Perspect 69:215–220.

Sax NT, Lewis RJ (1987) Hawley's condensed chemical dictionary, 11th Ed, Van Nostrand Reinhold, New York, NY. pp 261, 1175.

Silver EH, Kuttab SH, Hasen T, Hassan M (1982) Structural consideration in the metabolism of nitriles to cyanide *in vivo*. Drug Metab Dispos 10(5):495–498.

Simmon VF, Kaubanen K, Tardiff RG (1977) Mutagenic activity of chemicals identified in drinking water. In: Scott D, Bridges BA, Sobels FH (eds) Progress in genetic toxicology. Elsevier/North-Holland Biomedical Press, Amsterdam. pp 249–258.

Smith MK, Zenick H, George EL (1986). Reproductive toxicology of disinfection by-products. Envtl Health Perspect 69:177–182.

Smith MK, George EL, Zenick H., Manson JM, Stober JA (1987) Developmental toxicity of halogenated acetonitriles: drinking water by-products of chlorine disinfection. Toxicology 46:83–93.

Smith MK, Randall JL, Tocco DR, York RG, Stober JA, Read EJ (1988) Teratogenic effects of trichloroacetonitrde in the Long-Evans rat. Teratology 38:113–120.

Smith MK, Randall JL, Stober JA, Read EJ (1989) Developmental toxicity of dichloroacetonitrile: a by-product of drinking water disinfection. Fundam Appl Toxicol 2:765–772.

Stevens A, Moore L, Slocum C, Seeger D, Ireland J (1990) Byproducts of chlorination in ten operating utilities. In: Jolley RL, Condie LW, Johnson JD, Katz S, Minear R, Mattice JS, Jacobs VA (eds) Water chlorination: chemistry, environmental impact and health effect, Vol 6. Lewis Publishers, Chelsea, MI. pp 579–604.

Suffet IH, Brenner L, Cairo PR (1980) GC/MS identification of trace organics in Philadelphia drinking waters during a 2-year period. Water Res 14:853–867.

Swenberg J, Kerns W, Pavkov K, Mitchell R, Gralla EJ (1980) Carcinogenicity of formaldehyde vapor: interim findings in a long-term bioassay of rats and mice. Dev Toxicol Environ Sci 8:283–286.

Tanii H, Hashimoto K (1984) Studies on the mechanism of acute toxicity of nitriles in mice. Arch Toxicol 55:47–54.

Trehy ML, Bieber TL (1981) Detection, identification and quantitative analyses of dihaloacetonitriles in chlorinated natural waters. In: Keith LH (ed) Advances in the identification and analysis of organic pollutants in water, Vol 2. Ann Arbor Science Publishers, Ann Arbor, MI. pp 941–975.

Trehy ML, Yost RA, Miles CJ (1986) Chlorination byproducts of amino acids in natural waters. Environ Sci Technol 20:1117–1122.

Trehy ML, Yost RA, Miles CJ (1987) Aminoacids as potential precursors for halogenated byproducts formed in chlorination of natural waters. In: 194th American Chemical Society Meeting. New Orleans, LA, August 30–September 4.

Ullberg S (1977) The technique of whole-body autoradiography: cryosectioning of large specimens. In: Alvfeldt O (ed) Science tools, special issue on whole body autoradiography. LKB Instr J, Bromma, Sweden. pp 2–29.

U.S. EPA (Environmental Protection Agency) (1995) Methods for the Determination of Organic Compounds in Drinking Water, Supplement II (EPA/600/R-92/129).

U.S. Environmental Protection Agency (2006) National Primary Drinking Water Regulations: Stage 2 Disinfectants and Disinfection Byproducts Rule. Fed Regist 71:387–493.

Villanueva CM, Cantor KP, Cordier S, Jaakkola JJ, King WD, Lynch CF, Porru S, Kogevinas M (2004) Disinfection byproducts and bladder cancer: a pooled analysis. Epidemiology 15:357–367.

Waller K, Swan SH, DeLorenze G, Hopkins B (1998) Trihalomethanes in drinking water and spontaneous abortion. Epidemiology 9:134–40.

Weinberg HS, Krasner SW, Richardson SD, Thruston ADJ (2002) The Occurrence of Disinfection By-Products (DBPs) of Health Concern in Drinking Water: Results of a Nationwide DBP Occurrence Study; EPA/600/R02/068; U.S. EPA: Washington, DC.

Weisel CP, Kim H, Haltmeier P, Klotz JB (1999) Exposure estimates to disinfection by-products of chlorinated drinking water. Environ Health Perspect 107:103–110.

WHO (World Health Organization) (1993) Guidelines for drinking water quality 2nd Ed, Vol 1. World Health Organization, Geneva.

Williams DT, LeBel GL, Benoit FM (1997) Disinfection by-products in Canadian drinking water. Chemosphere 34:299–316.

Woo YT, Arcos JC, Lai DY (1988) Metabolic and chemical activation of carcinogens: an overview. In: Politzer P, Martin EJJ (eds) Chemical Carcinogens: Activation Mechanisms, Structural and Electronic Factors, and Reactivity. Elsevier, Amsterdam. pp 1–31.

Woo YT, Lai D, McLain JL, Manibusan MK, Dellarco V (2002) Use of mechanism-based structure-activity relationships analysis in carcinogenic potential ranking for drinking water disinfection by-products. Environ Health Perspect 110 (Suppl 1):75–87.

Wright JM, Schwartz J, Dockery DW (2004) The effect of disinfection byproducts and mutagenic activity on birth weight and gestational duration. Environ Health Perspect 112:920–925.

Yang X, Shang C, Westerhoff P (2007) Factors affecting formation of haloacetonitriles, haloketones, chloropicrin and cyanogen halides during chloramination. Water Res 41:1193–1200.

Zimmermann FK, Mohr A (1992) Formaldehyde, glyoxal, urethane, methyl carbamate, 2,3-butanedione, 2,3-hexanedione, ethyl acrylate, dibromoacetonitrile and 2-hydroxypropionitrile induce chromosome loss in Saccharomyces cerevisiae. Mutat Res 270(2):151–166.

Index

17ß-estradiol, vitellogenin induction in fish, 7

A
Abnormal gonadal development, fish, 7
Acute hemorrhagic cystitis, adenoviruses, 142
Acute respiratory disease, adenoviruses, 139
Acute toxicity, HANs (table), 176
Adenoviral pneumonias, acute respiratory disease, 140
Adenovirus disease outbreaks, waterborne, 147, 148
Adenovirus in children, studies (table), 144
Adenovirus serotypes, classification (table), 135
Adenovirus source, groundwater, 150
Adenovirus-associated disease, humans (table), 136
Adenoviruses and obesity, link, 141
Adenoviruses in water, survival, 148
Adenoviruses, acute hemorrhagic cystitis, 142
Adenoviruses, acute respiratory disease, 139
Adenoviruses, disinfection (table), 156
Adenoviruses, economic impact, 152
Adenoviruses, eye infections, 141
Adenoviruses, human gastroenteritis, 138
Adenoviruses, human respiratory infections, 139
Adenoviruses, human viral studies, 137
Adenoviruses, human-associated diseases, 135
Adenoviruses, impact on AIDS patients, 147
Adenoviruses, impact on children, 143
Adenoviruses, impact on immunocompromised, 146
Adenoviruses, in drinking water (table), 151
Adenoviruses, meningoencephalitis, 142
Adenoviruses, morbidity and mortality, 142
Adenoviruses, mortality ratios (table), 143
Adenoviruses, occurrence (table), 149
Adenoviruses, occurrence in sewage, 149
Adenoviruses, oncogenicity in humans, 138
Adenoviruses, pharyngoconjunctival fever, 140
Adenoviruses, removal from water, 155
Adenoviruses, risk assessment, 152
Adenoviruses, risks in drinking-recreational water (table), 153
Adenoviruses, risks of infection (table), 154
Adenoviruses, structure and properties, 135
Adenoviruses, waterborne disease, 147
Adenoviruses, waterborne organisms, 133 ff.
AIDS patients, impact of adenoviruses, 147
Alkylphenols, vitellogenin induction in fish, 7
Altered sex skew, chemical effects on birds, 27
Altered spermatogenesis in fish, 17ß-estradiol, 8
Amphibian deformities, PCBs, 18
Amphibian early life stage, chemical impact, 13
Amphibian population decline, pesticides, 17
Amphibian reproduction, PCB and pesticide effects, 14
Amphibians, chemical effects, 12, 18
Amphibians, estrogen effects of PCBs-DDT, 15
Amphibians, fertilizer impact, 17
Anthropogenic chemical impact, mammals, 30
Anthropogenic chemicals, effect on amphibians, 12
Anthropogenic chemicals, effects in reptiles, 19
Anthropogenic chemicals, impact on fish, 4
Anthropogenic chemicals, wildlife exposure, 2
Aquatic plant biomarker, chlorophyll fluorescence, 95
Aquatic plant biomarkers, flavonoids, 83
Aquatic plant biomarkers, pathway specific metabolites, 72
Aquatic plant biomarkers, Phase I metabolic enzymes, 75
Aquatic plant biomarkers, Phase II metabolic enzymes, 78
Aquatic plant biomarkers, phytochelatins, 81
Aquatic plant biomarkers, plant photosynthetic pigments, 92

Aquatic plant biomarkers, reactive oxygen species and scavenging enzymes, 89
Aquatic plant biomarkers, stress or heat- shock proteins, 85
Aquatic plants and biomarkers, concepts, 50
Aquatic plants and xenobiotic stress, biomarkers to measure (table), 53
Aquatic plants, biomarkers to detect xenobiotic stress (illus.), 69
Aquatic plants, selecting biomarkers, 49 ff.
Atrazine, effects in amphibians, 15

B

Biological activities compared, HANs (table), 193
Biomarkers in aquatic plants, measuring xenobiotic stress (table), 53
Biomarkers in aquatic plants, pathway specific metabolites, 72
Biomarkers in aquatic plants, responses, 50
Biomarkers of exposure, evaluating methylmercury, 122
Biomarkers, selection and utility, 49 ff.
Biomarkers, xenobiotic stress detection (illus.), 69
Bird deformities, organochlorine effects, 26
Bird reproduction, chemical effects, 24
Birds, oil spill effects, 28
Bottom-up approach, wildlife testing, 2

C

Carbaryl, effects in amphibians, 15
Carcinogenesis, HANs, 183, 186
Cartilaginous fish, chemicals and reproduction, 10
Chalcone synthase, quantifying flavonoids, 84
Chemical activities compared, HANs (table), 193
Chemical effects, amphibians, 18
Chemical impact, feral mammals, 30
Chemical impact, fish, 4
Chemical impact, vertebrate wildlife, 1 ff.
Chemical-induced altered sex skew, birds, 27
Chemicals, effects on cartilaginous fish, 10
Chemicals, impact on birds, 23
Chemicals, interaction with fish, 4
Chloramine, HANs link, 171
Chlorophyll fluorescence, applications to evaluate xenobiotic stress, 96
Chlorophyll fluorescence, aquatic plant biomarker, 95
Chondrichthyes, chemicals and reproduction, 10
Cycling of environmental mercury, sources, 112

D

DBPs (disinfection by-products), chlorine compounds, 169
DDT, altered sex skew in birds, 27
DDT, amphibian estrogenic-androgenic effects, 15
DDT, bird reproductive effects, 24
DDT, eggshell thinning, 28
DDT, fish effects, 5
Delayed sexual maturation in fish, mill effluents, 9
Developmental abnormalities in reptiles, organochlorine effects, 22
Developmental effects, HANs, 178
Dietary exposure to methylmercury, marine source, 115
Disinfection by-products (DBPs), chlorine compounds, 169
Disinfection, adenoviruses (table), 156
Drinking water source, adenoviruses (table), 151
Drinking water, HANs contaminants, 171
Drinking water, HANs residues (table), 172

E

Economic impact, adenoviruses, 152
EDC (endocrine disrupting chemicals) disruption, mammals, 31, 33, 34
EDC effects, reptile populations, 21
EDCs, wildlife vertebrates, 3
EDCs, bird reproductive effects, 24
EDCs, flatfish effects, 5
EDCs, gender determination effects, 20
EDCs, general fish effects, 11
EDCs, mimicry of natural hormones, 7
Effluent effect on fish, pulp-paper mills, 8
Eggshell thinning, DDT, 28
Endocrine disrupting chemicals (EDCs), wildlife vertebrates, 3
Endocrine disrupting properties, metals in fish, 10
Endocrine disruption, fish gonad effects, 8
Endocrine regulation in amphibians, EDC effects, 14
Environmental chemicals, impact on wildlife, 1 ff.
Environmental implications, mercury speciation, 114
Environmental mercury, global exposure and cycling, 112
Estrogen in amphibians, PCB and DDT effects, 15
Estrogenic substances, vitellogenin induction, 6

Evaluating xenobiotic stress, applications, 70
Eye infections, adenoviruses, 141

F
Feral mammals, chemical impact, 30
Fertilizer impact, amphibians, 17
Fish, methylmercury consumption risks and benefits, 121
Fish effects of metals, endocrine disrupting properties, 10
Fish effects, impact of heavy metals, 9
Fish effects, oil spills, 5
Fish gonadal effects, endocrine disruption, 8
Fish hormone disruption, vitellogenin induction, 6
Fish, alkylphenol and 17ß-estradiol effects, 7
Fish, anthropogenic chemical impact, 4
Fish, reproductive impairment, 5
Fish, response to chemical exposure, 4
Fish, source of methylmercury, 115
Flatfish effects, EDCs, 5
Flavonoids, applications to evaluate xenobiotic stress, 84
Flavonoids, aquatic plant biomarkers, 83
Flavonoids, quantifying with chalcone synthase, 84
Freshwater mammals, EDC disruption, 33

G
Gastroenteritis in humans, adenovirus induction, 138
Gastrointestinal tract injury (GIT), HANs, 181
Gender determination effects, insecticides, 20
Gender determination, EDC effects, 20
Gene expression biomarkers, quantification, 70
Gene expression, biomarkers and plants, 52
Gene expression, evaluating xenobiotic stress, 70
Genotoxicity, HANs, 183
Global cycling, mercury exposure, 112
Gonadal maldevelopment, fish, 7
Groundwater, adenovirus source, 150

H
Haloacetonitriles, metabolism and toxicity, 169 ff.
Halogenated acetonitriles (HANs), formation and prevalence, 170
HANs (halogenated acetonitriles), formation, 170
HANs GIT (gastrointestinal tract) injury, pathways (illus.), 182
HANs, acute toxicity (table), 176
HANs, carcinogenesis, 183
HANs, chloramine use, 171
HANs, drinking water residues (table), 172
HANs, gastrointestinal tract (GIT) injury, 181
HANs, genotoxic effects, 183
HANs, liver tumor production, 188
HANs, mechanistic pathways, 188
HANs, metabolic pathway (illus.), 174
HANs, oxidative stress implications, 190
HANs, pharmacokinetics and metabolism, 171
HANs, prevalence in drinking water, 171
HANs, relative biological-chemical activities, 192
HANs, reproductive and developmental toxicity, 178
HANs, skin and lung tumor production, 187
Health effects in humans, methylmercury, 111 ff.
Heat shock proteins, aquatic plant biomarkers, 85
Hormone effects in fish, vitellogenin induction, 6
Human adenovirus serotypes, classification (table), 135
Human adenoviruses, structure and properties, 135
Human adenoviruses, taxonomy, 134
Human diseases, adenovirus association, 135
Human exposure pathways, methylmercury, 114
Human exposure, mercury, 112
Human hair residues, mercury (table), 118
Human health effects, methylmercury exposure, 111 ff.
Human illnesses, adenoviruses (table), 136
Human metabolism, methylmercury, 119
Human reports, mercury exposure, 116
Humans and weak immunity, adenovirus impact, 146

I
Impact of chemicals, amphibians, 12
Impact of chemicals, mammals, 30
Impact of fertilizers, amphibians, 17
Impact of heavy metals, fish effects, 9
Impact on birds, chemicals, 23
Impact on children, adenoviruses, 143
Impact on fish, pulp-paper mill effluents, 8
Impact on reptiles, chemicals, 19
Impact on vertebrates, environmental chemicals, 1 ff.

Infection risks, adenoviruses (table), 154
Infections in humans, adenoviruses (table), 136
Insecticide effects, gender determination, 20

L
Liver tumor production, HANs, 188
Lung tumor production, HANs, 187

M
Mammalian toxicity, HANs, 176
Mammals, chemical impact, 30
Mammals, EDC disruption, 31, 33, 34
Marine mammals, EDC disruption, 31
Marine sources, methylmercury exposure, 115
Mechanism of action, HANs, 188
Mechanistic pathways, HANs, 188
Meningoencephalitis, adenoviruses, 142
Mercury exposure, human reports, 116
Mercury exposure, humans, 112
Mercury exposure, sources and cycling, 112
Mercury levels, human hair (table), 118
Mercury speciation, environmental effects, 114
Metabolic enzymes, aquatic plant biomarkers, 75, 78
Metabolic pathway, HANs (illus.), 174
Metabolism, haloacetonitriles, 169 ff.
Metabolism, HANs, 174
Metabolism, methylmercury, 119
Metals, fish effects, 9
Methylmercury and fish consumption, risks and benefits, 121
Methylmercury exposure, human health effects, 111 ff.
Methylmercury exposure, human pathways, 114
Methylmercury exposure, marine food source, 115
Methylmercury poisoning, Minamata Bay incident, 115
Methylmercury, exposure biomarkers, 122
Methylmercury, human toxicity, 119
Methylmercury, metabolism, 119
Methylmercury, risk evaluation, 120
Methylmercury, toxicokinetics, 117
Methylmercury, transport and mobility, 124
Military institutions, acute respiratory disease, 140
Minamata Bay, methylmercury incident, 115
Mobility, methylmercury, 124
Morbidity in humans, adenoviruses, 142
Mortality in humans, adenoviruses, 142
Mortality ratios, adenoviruses (table), 143

O
Obesity and adenoviruses, link, 141
Oil spills, effect on birds, 28
Oil spills, fish effects, 5
Oncogenicity, adenoviruses, 138
Organochlorine chemicals, abnormalities in reptiles, 22
Organochlorine chemicals, effects in fish, 5
Organochlorine insecticide effects, reptile populations, 21
Organochlorines, altered sex skew in birds, 27
Organochlorines, bird deformities role, 26
Organochlorines, bird reproductive effects, 24
Organochlorines, effects on cartilaginous fish, 10
Organochlorines, impact on birds, 23
Organochlorines, shark effects, 11
Osteichthyes, reproductive parameters, 4
Oxidative stress, HANs adverse effects, 190

P
PAHs (polyaromatic hydrocarbons), fish effects, 5
Paper-pulp mill effluents, impact on fish, 8
Pathway specific metabolites, aquatic plant biomarkers, 72
Pathway specific metabolites, quantification, 73
PCBs (polychlorinated biphenyls), fish effects, 5
PCBs, amphibian deformities, 18
PCBs, amphibian estrogenic-androgenic effects, 15
PCBs, bird reproductive effects, 24
PCBs, effects in amphibians, 14
PCBs, endocrine disruption, 33, 34
PCBs, fish effects, 9
PCBs, reptile population effects, 21
Persistent organic pollutants (POPs), bird effects, 23
Pesticides, effects in amphibians, 14
Pharmacokinetics, HANs, 171
Pharyngoconjunctival fever, adenoviruses, 140
Phase I enzymes, applications to evaluate xenobiotic stress, 77
Phase I metabolic enzymes, aquatic plant biomarkers, 75
Phase I metabolic enzymes, quantifying xenobiotic stress, 76
Phase II enzymes, applications to evaluate xenobiotic stress, 79
Phase II metabolic enzymes, aquatic plant biomarkers, 78

Phase II metabolic enzymes, quantifying xenobiotic stress, 79
Phytochelatins, applications to evaluate xenobiotic stress, 82
Phytochelatins, aquatic plant biomarkers, 81
Phytochelatins, quantifying xenobiotic stress, 82
Phytoestrogen fish exposure, mill effluents, 9
Plant photosynthetic pigments, applications to evaluate xenobiotic stress, 93
Plant photosynthetic pigments, aquatic plant biomarkers, 92
Poisoning in Iraq, methyl- and ethyl-mercury, 115
Polyaromatic hydrocarbons (PAHs), fish effects, 5
Polychlorinated biphenyls (PCBs), fish effects, 5
POPs (persistent organic pollutants, bird effects, 23
Prevalence, HANs in drinking water, 171
Properties, human adenoviruses, 135
Pulp-paper mill effluents, impact on fish, 8

Q

Quantification, chlorophyll fluorescence, 96
Quantification, pathway specific metabolites, 73
Quantification, Phase I metabolic enzymes, 76
Quantification, Phase II metabolic enzymes, 79
Quantification, phytochelatins 82
Quantification, plant photosynthetic pigments, 93
Quantification, reactive oxygen species, 89
Quantification, scavenging enzymes, 90
Quantification, stress proteins, 86
Quantifying stress, gene expression biomarkers, 70

R

Reactive oxygen species, applications to evaluate xenobiotic stress, 91
Reactive oxygen species, aquatic plant biomarkers, 89
Reproduction in amphibians, chemical effects, 12
Reproductive effects in amphibians, atrazine, 15
Reproductive effects, pesticides and amphibians, 13
Reproductive impact of steroid hormones, birds, 26
Reproductive impairment, in fish, 5
Reproductive parameters, *Osteichthyes*, 4
Reproductive toxicity, HANs, 178
Reptile populations, EDC and insecticide effects, 21
Reptile effects, chemicals, 19
Respiratory infections, human adenoviruses, 139
Risk assessment, adenoviruses, 152
Risk evaluation, methylmercury, 120
Risks and benefits, methylmercury intake in fish, 121
Risks of adenoviruses, drinking-recreational water (table), 153

S

Scavenging enzymes, aquatic plant biomarkers, 89
Sewage sources, adenoviruses (table), 149
Sewage treatment discharges, fish effects, 7
Sewage, source of adenoviruses, 148
Shark effects, organochlorines, 11
Skin tumor production, HANs, 187
Spermatogenesis alterations, 17ß-estradiol, 8
Steroid hormone effects, birds, 26
Stingray effects, tributyltin and its oxide, 11
Stress proteins, applications to evaluate xenobiotic stress, 87
Stress proteins, aquatic plant biomarkers, 85
Surface water sources, adenoviruses (table), 149

T

Terrestrial mammals, endocrine disruption, 34
Testicular growth effects, EDCs, 7, 8
Top-down approach, wildlife testing, 2
Toxicity in humans, methylmercury, 119
Toxicity of HANs, reproductive and developmental, 178
Toxicity, haloacetonitriles, 169 ff.
Toxicity, HANs, 176
Toxicokinetics, methylmercury, 117
Tributyltin oxide, stingray effects, 11
Tributyltin, stingray effects, 11

V

Viral studies, adenoviruses in humans, 137
Vitellogenin induction in fish, alkylphenols and
Vitellogenin induction, fish hormone effects, 6

W

Wastewater, fish vitellogenin induction, 7
Water pollution, fish vitellogenin induction, 7
Water treatment, to remove adenoviruses, 155
Waterborne disease outbreaks, adenoviruses, 147
Waterborne organisms, adenoviruses, 133 ff.
Wildlife testing approaches bottom-up or top-down (illus.), 3
Wildlife testing top-down or bottom-up approaches, 2
Wildlife vertebrates, chemical impact, 1ff.

X

Xenobiotic stress detection, biomarkers (illus.), 69
Xenobiotic stress evaluation, chlorophyll fluorescence, 96
Xenobiotic stress evaluation, Phase I enzymes, 77
Xenobiotic stress evaluation, plant photosynthetic pigments, 93
Xenobiotic stress evaluation, reactive oxygen species, 91
Xenobiotic stress evaluation, stress or heat-shock proteins, 87
Xenobiotic stress, biomarker measurement in aquatic plants (table), 53
Xenobiotic stress, pathway specific metabolite applications, 73